T0212782

Lecture Notes in Computer Science 9288

Commenced Publication in 1973
Founding and Former Series Editors:
Gerhard Goos, Juris Hartmanis, and Jan van Leeuwen

Editorial Board

David Hutchison
  Lancaster University, Lancaster, UK
Takeo Kanade
  Carnegie Mellon University, Pittsburgh, PA, USA
Josef Kittler
  University of Surrey, Guildford, UK
Jon M. Kleinberg
  Cornell University, Ithaca, NY, USA
Friedemann Mattern
  ETH Zurich, Zürich, Switzerland
John C. Mitchell
  Stanford University, Stanford, CA, USA
Moni Naor
  Weizmann Institute of Science, Rehovot, Israel
C. Pandu Rangan
  Indian Institute of Technology, Madras, India
Bernhard Steffen
  TU Dortmund University, Dortmund, Germany
Demetri Terzopoulos
  University of California, Los Angeles, CA, USA
Doug Tygar
  University of California, Berkeley, CA, USA
Gerhard Weikum
  Max Planck Institute for Informatics, Saarbrücken, Germany

More information about this series at http://www.springer.com/series/7407

Jérôme Durand-Lose · Benedek Nagy (Eds.)

Machines, Computations, and Universality

7th International Conference, MCU 2015
Famagusta, North Cyprus, September 9–11, 2015
Proceedings

 Springer

Editors
Jérôme Durand-Lose
Université d'Orléans
Orléans
France

Benedek Nagy
Eastern Mediterranean University
Famagusta
North Cyprus

and

University of Debrecen
Debrecen
Hungary

ISSN 0302-9743   ISSN 1611-3349 (electronic)
Lecture Notes in Computer Science
ISBN 978-3-319-23110-5   ISBN 978-3-319-23111-2 (eBook)
DOI 10.1007/978-3-319-23111-2

Library of Congress Control Number: 2015946770

LNCS Sublibrary: SL1 – Theoretical Computer Science and General Issues

Springer Cham Heidelberg New York Dordrecht London

© Springer International Publishing Switzerland 2015
This work is subject to copyright. All rights are reserved by the Publisher, whether the whole or part of the material is concerned, specifically the rights of translation, reprinting, reuse of illustrations, recitation, broadcasting, reproduction on microfilms or in any other physical way, and transmission or information storage and retrieval, electronic adaptation, computer software, or by similar or dissimilar methodology now known or hereafter developed.
The use of general descriptive names, registered names, trademarks, service marks, etc. in this publication does not imply, even in the absence of a specific statement, that such names are exempt from the relevant protective laws and regulations and therefore free for general use.
The publisher, the authors and the editors are safe to assume that the advice and information in this book are believed to be true and accurate at the date of publication. Neither the publisher nor the authors or the editors give a warranty, express or implied, with respect to the material contained herein or for any errors or omissions that may have been made.

Printed on acid-free paper

Springer International Publishing AG Switzerland is part of Springer Science+Business Media
(www.springer.com)

Preface

This volume contains the papers presented at the 7th international conference on Machines, Computations and Universality (MCU 2015) held during September 9–11, 2015, in Famagusta.

MCU explores computation in the setting of various discrete models (Turing machines, register machines, cellular automata, tile assembly systems, rewriting systems, molecular computing models, neural models…) and analog and hybrid models (BSS machines, infinite time cellular automata, real machines, quantum computing…). There is a particular, but not exclusive, emphasis given to:

- The search for frontiers between decidability and undecidability in the various models
- The search for the simplest universal models
- The computational complexity of predicting the evolution of computations in the various models
- The connection between parallelism and decidability, complexity and universality
- Universality and undecidability in continuous models of computation

Initiated by Maurice Margenstern, the MCU international conference series traces its roots back to the mid-1990s, and has since been concerned with gaining a deeper understanding of computation through the study of models of general purpose computation. Previous MCU conferences took place in Zürich, Switzerland (2013), Orléans, France (2007), Saint-Petersburg, Russia (2004), Chişinău, Moldova (2001), Metz, France (1998), and Paris, France (1995).

There were 23 papers submitted to the 2015 edition of MCU. Each submission was reviewed by four Program Committee members or sub-reviewers. The committee decided to accept ten papers. The program also included four invited talks. Various topics of MCU were overviewed by the invited talks:

- Jetty Klein from Leiden University is an expert on concurrency and Petri nets. In her talk she presented on-going research about generalized traces, i.e., an extension of Mazurkiewicz's trace theory.
- Linqiang Pan from Huazhong University of Science and Technology is working on membrane computing. His speciality is spiking neural P systems. In his talk he explained relationships and differences between spiking neural networks and spiking neural P systems, some classic and recent results.
- Anne Siegel is a senior researcher at CNRS, IRISA, Rennes. She is one of the leading scientists in qualitative dynamical systems in computer science, biology, and discrete mathematics. Her speech was about decidability and undecidability results connected to Rauzy fractals.
- Mike Stannett, from the University of Sheffield, is a leading expert on hypercomputations. He explained that the computational power of a distributed system could depend on the physical environment, i.e., on spacetime geometry in which the computation is performed.

Informal presentations were also made at the conference but they are not mentioned in these proceedings.

MCU 2015 was also an occasion to see the historical city of Famagusta and some of the beautiful surroundings of the city and enjoy the Mediterranean Sea. The city of Famagusta is one of the finest examples of medieval well-preserved architecture in the eastern Mediterranean. The ruins of the ancient city of Salamis are one of the most impressive archaeological sites on the island.

The EasyChair conference system[1] was used to handle submissions, the review process, and LNCS production. As usual, it worked smoothly and perfectly. We definitely recommend it.

MCU 2015 was organized by the Eastern Mediterranean University[2], Famagusta, North Cyprus. We are very thankful to the Eastern Mediterranean University for making MCU possible and successful. We would like to thank Necdet Osam, Rector of Eastern Mediterranean University, for his energetic support, and Ahmet Sözen, Vice Rector for Academic Affairs of Eastern Mediterranean University, for helping us to resolve organizational problems.

The MCU 2015 organizing team did a great job. As far as the Organizing Committee is concerned, our first very warm thanks go to Rza Bashirov, Chair of MCU 2015 Organizing Committee, who kept track of every stage of event planning for MCU 2015, striving to reach the highest quality in organizational actions. Special thanks go to Mehmet Bozer, who carefully selected prizes for the best paper and best student paper awards as well as gadgets/giveaways for the conference bags. We also thank Hakan Arslan, who on behalf of the web office in Eastern Mediterranean University created the nicest website for MCU 2015. We thank in particular Hasan Arslan, the faculty administrator at the Faculty of Arts and Sciences, who performed his task with great professionalism, helping in the event's accounting and financial management. We thank Selda Adaöz and Gizem Sarca from EMU Press for putting scientific things in an artistic way: We had the most beautiful announcement poster for the conference. Finally, we extend our sincere appreciation to our sponsors:

– Deniz Plaza[3] for providing conference bags
– Laboratoire d'Informatique Fondamentale d'Orléans (LIFO)[4] for partial support in financial aspects

The editors warmly thank the Program Committee, the invited speakers, the authors of the papers, the external reviewers, the speakers of informal presentations, and all the participants for their contribution to the success of the conference.

July 2015   Jérôme Durand-Lose
                                                         Benedek Nagy

---

[1] http://www.easychair.org/

[2] http://www.emu.edu.tr/

[3] http://www.denizplaza.net/

[4] http://www.univ-orleans.fr/lifo/?lang=en

Organization

Program Committee

Andrew Adamatzky	University of the West of England, UK
Rza Bashirov	Eastern Mediterranean University, Famagusta
Laurent Bienvenu	CNRS, Université Paris Diderot, France
Erzsébet Csuhaj-Varjú	Eötvös Loránd University, Hungary
Jérôme Durand-Lose	University of Orléans, France (Co-chair)
Henning Fernau	University of Trier, Germany
Rudolf Freund	University of Vienna, Austria
Gabriel Istrate	West University of Timisoara, Romania
Jarkko Kari	University of Turku, Finland
Martin Kutrib	Universität Giessen, Germany
Peter Leupold	University of Leipzig, Germany
Maurice Margenstern	University of Lorraine, France
Kenichi Morita	Hiroshima University, Japan
Benedek Nagy	Eastern Mediterranean University (Co-chair) and University of Debrecen, Hungary
Turlough Neary	University of Zürich and ETH Zürich, Switzerland
Matthew Patitz	University of Arkansas, USA
Gheorghe Păun	Romanian Academy, Bucharest, Romania
Igor Potapov	University of Liverpool, UK
K.G. Subramanian	University of Science, Malaysia
Klaus Sutner	University Carnegy-Mellon, USA
György Vaszil	University of Debrecen, Hungary
Sergey Verlan	University of Paris Est, France

Steering Committee

Matthew Cook	University of Zürich and ETH Zürich, Switzerland
Erzsébet Csuhaj-Varjú	Eötvös Loránd University, Budapest, Hungary
Jérôme Durand-Lose	University of Orléans, France (Chair)
Natasha Jonoska	University of South Florida, USA
Maurice Margenstern	University of Metz, France (Honorary Chair)
Kenichi Morita	Hiroshima University, Japan
Gheorghe Păun	Romanian Academy, Bucharest, Romania
Arto Salomaa	University of Turku, Finland
K.G. Subramanian	University of Science, Malaysia

Organizing Committee Chairs

Rza Bashirov
Benedek Nagy

Invited Talks

Decidability Problems for Self-induced Systems Generated by a Substitution

Timo Jolivet[1] and Anne Siegel[2,3]

[1] Université Paul Sabatier, Institut de mathmatique de Toulouse,
Toulouse, France
[2] CNRS, Université de Rennes 1, IRISA-UMR 6074, Rennes, France
anne.siegel@irisa.fr
[3] Inria Rennes - Bretagne Atlantique, Team Dyliss, Rennes, France

Abstract. In this talk we will survey several decidability and undecidability results on topological properties of self-affine or self-similar fractal tiles. Such tiles are obtained as fixed point of set equations governed by a graph. The study of their topological properties is known to be complex in general: we will illustrate this by undecidability results on tiles generated by multitape automata. In contrast, the class of self affine tiles called Rauzy fractals is particularly interesting. Such fractals provide geometrical representations of self-induced mathematical processes. They are associated to one-dimensional combinatorial substitutions (or iterated morphisms). They are somehow ubiquitous as self-replication processes appear naturally in several fields of mathematics. We will survey the main decidable topological properties of these specific Rauzy fractals and detail how the arithmetic properties of the substitution underlying the fractal construction make these properties decidable. We will end up this talk by discussing new questions arising in relation with continued fraction algorithm and fractal tiles generated by S-adic expansion systems.

Concurrency, Histories, and Nets

Jetty Kleijn

LIACS, Leiden University, P.O.Box 9512
NL-2300 RA Leiden, The Netherlands
h.c.m.kleijn@liacs.leidenuniv.nl

Abstract

In the setting of discrete sequential system models such as finite automata and Turing machines, computations can be abstracted to sequences of symbols (words). Each symbol stands for an atomic action and each symbol occurrence represents an execution of the corresponding action.

In a concurrent system however, actions are not necessarily executed one after the other and sequential behavioural descriptions such as words, lack information on possible concurrency of events. In particular, no distinction can be made between necessary orderings of causally related events and observational, accidental, orderings. Still, also for concurrent systems, one may want to opt for an abstract language semantics to describe the evolutions of a system.

To distinguish between causally related and independent events, the, by now classical, Mazurkiewicz trace approach extends the alphabet of action symbols with a binary relation providing information on the independence of actions. Then all words that differ only in the ordering of occurrences of independent actions are identified as representing observations of the same concurrent computation of the system. The resulting equivalence classes are referred to as traces. The causal relations between symbol occurrences are common to (invariant among) all words constituting a trace and form an acyclic dependence graph that identifies the trace. The transitive closure of the dependence graph describes the underlying invariant causality structure of the trace as a labelled partial order.

All this leads to an order-theoretic counterpart of the trace approach within which concurrent behaviour is on the one hand represented as a history, i.e., an invariant closed set of (labelled) total orders, and on the other hand through a (labelled) partial order. These descriptions are in one-to-one correspondence because a history can be obtained by linearising (saturating) a partial order description in all possible ways, and a partial order description of concurrent behaviour is derived from the associated history by intersecting its total orders.

Finally, Elementary Net (EN) systems, generally regarded as the most fundamental class of Petri Nets, are the operational model that inspired the introduction of traces. They can be seen as concurrent counterparts of finite automata with their underlying net structure defining an independence relation over the transitions (actions) of the system. Their behaviour is in accordance with the 'true concurrency' paradigm formalised in the trace approach, that equates independence and lack of ordering.

There are however aspects of concurrency that cannot be modeled adequately in terms of partial orders alone and hence also not by classical traces. This is in particular the case when in a concurrent computation, actions cannot only be reported as occurring one after the other, but may be registered also as occurring simultaneously. Then concurrent computations are represented in terms of step sequences, i.e., sequences of sets of (one or more) actions recorded as occurring simultaneously. As before in the case of words, such sequences lack precise information on the relationships between events: no distinction is made between necessary orderings and simultaneity of events and between observational ordering and grouping. Moreover, to capture the underlying invariants of a concurrent computation described using steps, one needs relations like 'before or in the same step' and 'unordered but not in the same step' which cannot be expressed using partial orders as these would identify independence, unorderedness, and now also simultaneity.

In a theory of step traces that extends Mazurkiewicz' approach to step sequences, the alphabet of action symbols has three binary relations providing information on the relationships between actions. These fundamental relations are (*i*) simultaneity — indicating that actions can occur simultaneously and defining the legal steps of the system; (*ii*) serialisability — indicating a possible execution order of two potentially simultaneous actions and making it possible to split and combine steps; (*iii*) interleaving — indicating that actions cannot occur simultaneously though no specific order is required, and making it possible to swap steps on basis of interleaving and serialisability. Together, they form the basis of the identification of step sequences describing the same concurrent computation. The resulting equivalence classes of step sequences are referred to as step traces. Moreover, the clear semantical meaning of the three relations makes it possible to distinguish in a natural way eight subclasses of extended concurrency alphabets defining specific types of traces including one corresponding to the original traces and others already known from the literature as extensions.

The (labelled) relational structures that describe the invariant, common relationships between the symbol occurrences in the step sequences forming a trace, have a weak causality ('not later than') and a mutex relation ('not ordered and not simultaneous'). The resulting dependence structures satisfy a form of acyclicity and can be closed to yield a so-called invariant structure which identifies the step trace. The saturated extensions of invariant structures correspond to step sequences in the same way as total orders correspond to words. Moreover, all step sequences corresponding to saturated extensions of an invariant structure belong to the same step trace. As each invariant structure is the intersection of its saturated extensions, concurrent behaviour can again be represented by histories, now (invariant closed) sets of saturated invariant structures. Invariant structures turn out to be the most general relational structures in the sense that they can capture any history. It follows that step traces are the most general version of Mazurkiewicz traces in the context of step sequences. The various subclasses of step traces have correspondingly simplified invariant structures.

The EN systems that inspired Mazurkiewicz' approach have a sequential execution semantics. In the now extended framework, such traces are defined by the subclass of concurrency alphabets with an empty simultaneity relation. EN systems with a step semantics (under which transitions can occur simultaneously) define step traces subject to the concurrency paradigm by which lack of ordering is the same as simultaneity.

In this case, concurrency alphabets have an empty interleaving relation and the simultaneity and serialisability relations are the same, both coinciding with the independence of transitions and thus defined on basis of structural information obtained from the net. For EN systems extended with inhibitor arcs and operating under the step semantics, unorderedness of transition occurrences implies their simultaneity but the converse need not be the case. Here we deal with (the already known) comtraces with concurrency alphabets with an empty interleaving relation. Finally, to capture the most general case when there is no assumed relation between unorderedness and simultaneity, EN systems with inhibitor arcs can be further extended with mutex arcs that prohibit the simultaneous execution of otherwise independent transitions. The resulting ENIM systems thus fit the least restrictive concurrency paradigm for histories relating to step sequences and can therefore be viewed as the most general EN systems model.

Acknowledgement. This survey is based on ongoing research carried out together with Ryszard Janicki (McMaster University, Hamilton, Canada), Maciej Koutny (Newcastle University, Newcastle upon Tyne, UK), and Łukasz Mikulski (Nicolaus Copernicus University, Toru, Poland).

Literature

Mazurkiewicz, A..: Basic notions of trace theory. In: de Bakker, J.W., de Roever, W.-P., Rozenberg, G. (eds.) Linear Time, Branching Time and Partial Order in Logics and Models for Concurrency. LNCS, vol. 354, pp. 285–363. Springer, Heidelberg (1989)

Diekert, V., Rozenberg, G. (eds.): The Book of Traces. World Scientific (1995)

Kleijn, J., Koutny, M.: Mutex causality in processes and traces of general elementary nets. Fundam. Inform. **122** (1–2), 119–146 (2013)

Janicki, R., Kleijn, J., Koutny, M., Mikulski, Ł.: Characterising concurrent histories. Fundam. Inform. **139**, 21–42 (2015)

Janicki, R.., Kleijn, J., Koutny, M., Mikulski, Ł.: Generalising traces; TR-CS 1436. Newcastle University (2014)

Janicki, R., Kleijn, J., Koutny, M., Mikulski, Ł.: Step traces; Acta Inf. (to appear)

Janicki, R., Kleijn, J., Koutny, M., Mikulski, Ł.: Order structures for subclasses of generalised traces. In: Dediu, A.-H., Formenti, E., Martín-Vide, C., Truthe, B. (eds.) LATA 2015. LNCS, vol. 8977, pp. 689–700. Springer, Heidelberg (2015)

Spiking Neural P Systems: A Class of Parallel Computing Models Inspired by Neurons

Linqiang Pan

Key Laboratory of Image Processing and Intelligent Control,
School of Automation, Huazhong University of Science and Technology,
Wuhan 430074, Hubei, China
lqpan@mail.hust.edu.cn; lqpanhust@gmail.com

An Extended Abstract

Nature or natural systems are a rich source for the inspiration of new computational paradigms and techniques. Examples of nature inspired computational paradigms include evolutionary algorithms, artificial neural networks, colony optimization, swarm intelligence, simulated annealing, cellular automata, etc. Such research leads to various algorithms, even the design of novel computing systems that use natural media to compute.

Membrane computing is a computational paradigm inspired by cells, which aims to abstract computing ideas (data structures, operations with data, computing models, etc.) from the structure and the functioning of a single cell or from complexes of cells, such as tissues and organs including the brain [18, 20]. The obtained models are distributed and parallel computing devices, called P systems. In this talk, we consider a class of P systems inspired by neurons, called spiking neural P systems (SN P systems, for short) [11].

An SN P system consists of a set of neurons, which are placed in the nodes of a directed graph whose arcs represent the synapses. Each neuron can contain a number of copies of a single object type (called spike), spiking rules and forgetting rules. Using its rules, a neuron can send information (in the form of spikes) to all neurons connected by an outgoing synapse from it. The applicability of a rule is usually determined by checking the total number of spikes contained in the neuron against a regular expression associated with the rule. One of the neurons is the output neuron and its spikes are sent to the environment. A result can be associated with a computation in various ways: for example, as the time elapsed between the first two consecutive spikes sent to the environment by the system, that is, the time dimension is used as data support.

The SN P systems are a class of distributed parallel computing models in the framework of membrane computing, which have the same origin as recurrent neural networks, but the computing units (i.e., neurons) evolve in a quite different way. They use individual spikes, instead of an averaging mechanism like rate coding, allowing us to incorporate spatial and temporal information in computation, where the number and timing of spikes matter. In the above sense, SN P systems fall into the third generation of neural network models [13].

SN P systems are computationally powerful in the sense that only a small number of computing units (i.e., neurons) suffice for SN P systems to achieve universality (e.g., 84 or 49 neurons in [21], even 9 computing units arranged in a linear structure in the variant of SN P system, axon P system [30]). However, for recurrent neural networks, one of the most investigated models of artificial neural networks, it was proved that 886 computing units are enough for this device to achieve universality for computing functions [24].

With various mathematical or biology-inspired motivation, many variants of SN P systems were introduced, for example, SN P systems with anti-spikes [14], SN P systems with astrocytes [1, 16, 19], SN P systems with weighted synapses [17], SN P systems with decaying spikes and/or total spiking [8], SN P systems with rules on synapses [25].

SN P systems can be used as number generating or accepting devices [10, 11, 26, 31], language generators [6, 7, 32], and function computing devices [21, 22, 33]. SN P systems can be also used to (theoretically) solve computationally hard problems in a feasible time [12, 15], even real-life problems, for example, representing and processing fuzzy and uncertain knowledge [28, 29], fuzzy inference and learning [27], fault diagnosis [23].

Some simulating/implementing tools of SN P systems were developed [2–5, 9].

In general, with the rich theoretical results and the initial application of SN P systems, it is expected to solve real-life problems by SN P systems.

References

1. Binder, A., Freund, R., Oswald, M., et al.: Extended spiking neural P systems with excitatory and inhibitory astrocytes. In: Proceedings of Fifth Brainstorming Week on Membrane Computing, pp. 63–72 (2007)
2. Cabarle, F., Adorna, H., Martinez-Del-Amor, M.A., et al.: Improving GPU simulations of spiking neural P systems. Rom. J. Inf. Sci. Tech. **15**(1), 5–20 (2012)
3. Cabarle, F., Adorna, H., Martinez-del-Amor, M.A.: Simulating spiking neural P systems without delays using GPUs. No. arXiv: 1104.1824 (2011)
4. Cabarle, F., Adorna, H., Martinez-del-Amor, M.A.: A spiking neural P system simulator based on CUDA. In: Gheorghe, M., Păun, G., Rozenberg, G., Salomaa, A., Verlan, S. (eds.) CMC 2011. LNCS, vol. 7184, pp. 87–103. Springer, Heidelberg (2012)
5. Cavaliere, M., Mura, I.: Experiments on the reliability of stochastic spiking neural P systems. Nat. Comput. **7**(4), 453–470 (2008)
6. Chen, H., Freund, R., Ionescu, M., Păun, Gh., Pérez-Jiménez, M.J.: On string languages generated by spikng neural P systems. Fund. Inform. **75**, 141–162 (2007)
7. Chen, H., Ionescu, M., Ishdorj, T.-O., Păun, A., Păun, Gh., Pérez-Jiménez, M.J.: Spiking neural P systems with extended rules: universality and languages. Nat. Comput. **7**(2), 147–166 (2008)
8. Freund, R., Ionescu, M., Oswald, M.: Extended spiking neural P systems with decaying spikes and/or total spiking. Int. J. Found. Comput. S. **19**(05), 1223–1234 (2008)
9. Gutiérrez-Naranjo, M.A., Pérez-Jiménez, M.J., Ramírez-Martínez, D.: A software tool for verification of spiking neural P systems. Nat. Comput. **7**(4), 485–497 (2008)

10. Ibarra, O.H., Păun, A., Rodríguez-Patón, A.: Sequential SNP systems based on min/max spike number. Theor. Comput. Sci. **410**(30–32), 2982–2991 (2009)
11. Ionescu, M., Păun, Gh., Yokomori, T.: Spiking neural P systems. Fund. Inform. **71**(2–3), 279–308 (2006)
12. Ishdorj, T.-O., Leporati, A., Pan, L., Zeng, X., Zhang, X.: Deterministic solutions to QSAT and Q3SAT by spiking neural P systems with pre-computed resources. Theor. Comput. Sci. **411**(25), 2345–2358 (2010)
13. Maass, W.: Networks of spiking neurons: the third generation of neural network models. Neural Netw. **10**(9), 1659–1671 (1997)
14. Pan, L., Păun, Gh.: Spiking neural P systems with anti-spikes. Int. J. Comput. Commun. Control **IV** (3), 273–282 (2009)
15. Pan, L., Păun, Gh., Pérez-Jiménez, M.J.: Spiking neural P systems with neuron division and budding. Sci. China Inform. Sci. **54**(8), 1596–1607 (2011)
16. Pan, L., Wang, J., Hoogeboom, H.J.: Spiking neural P systems with astrocytes. Neural Comput. **24**(3), 805–825 (2012)
17. Pan, L., Zeng, X., Zhang, X., et al.: Spiking neural P systems with weighted synapses. Neural Process. Lett. **35**(1), 13–27 (2012)
18. Păun, Gh.: Computing with membranes, J. Comput. Syst. Sci. **61**(1), 108–143 (2000)
19. Păun, Gh.: Spiking neural P systems with astrocyte-like control. J. Univers. Comput. Sci. **13** (11), 1707–1721 (2007)
20. Păun, Gh., Rozenberg, G., Salomaa, A. (eds.): Handbook of Membrane Computing. Oxford University Press, Cambridge (2010)
21. Păun, A., Păun, Gh.: Small universal spiking neural P systems. BioSystems **90**(1), 48–60 (2007)
22. Păun, A., Sidoroff, M.: Sequentiality induced by spike number in SNP systems: small universal machines. In: Gheorghe, M., Păun, Gh., Rozenberg, G., Salomaa, A., Verlan, S. (eds.) CMC 2011. LNCS, vol. 7184, pp. 333–345. Springer, Heidelberg (2012)
23. Peng, H., Wang, J., Pérez-Jiménez, M.J., et al.: Fuzzy reasoning spiking neural P system for fault diagnosis. Inform. Sci. **235**, 106–116 (2013)
24. Siegelmann, H.T., Sontag, E.D.: On the computational power of neural nets. J. Comput. Syst. Sci. **50**(1), 132–150 (1995)
25. Song, T., Pan, L., Păun, Gh.: Spiking neural P systems with rules on synapses. Theor. Comput. Sci. **529**, 82–95 (2014)
26. Wang, J., Hoogeboom, H.J., Pan, L., Păun, Gh., Pérez-Jiménez, M.J.: Spiking neural P systems with weights. Neural Comput. **22**(10), 2615–2646 (2010)
27. Wang, J., Peng, H.: Adaptive fuzzy spiking neural P systems for fuzzy inference and learning. Int. J. Comput. Math. **90**(4), 857–868 (2013)
28. Wang, J., Shi, P., Peng, H., et al.: Weighted fuzzy spiking neural P systems. IEEE Trans. Fuzzy Syst. **21**(2), 209–220 (2013)
29. Wang, J., Zhou, L., Peng, H., et al.: An extended spiking neural P system for fuzzy knowledge representation. Int. J. Innov. Comput. I. **7**(7), 3709–3724 (2011)
30. Zhang, X., Pan, L., Păun A.: On the Universality of Axon P Systems. IEEE Trans. Neur. Net. Lear. doi: 10.1109/TNNLS.2015.2396940.
31. Song, T., Pan, L., Păun, Gh.: Asynchronous spiking neural P systems with local synchronization. Inform. Sci. **219**, 197–207 (2013)
32. Zhang, X., Zeng, X., Pan, L.: On languages generated by asynchronous spiking neural P systems. Theor. Comput. Sci. **410**(26), 2478–2488 (2009)
33. Zhang, X., Zeng, X., Pan, L.: Smaller universal spiking neural P systems. Fund. Inform. **87** (1), 117–136 (2008)

Towards Formal Verification of Computations and Hypercomputations in Relativistic Physics

Mike Stannett

Department of Computer Science, University of Sheffield
Regent Court, 211 Portobello, Sheffield S1 4DP, UK
m.stannett@sheffield.ac.uk

Abstract. It is now more than 15 years since Copeland and Proudfoot introduced the term *hypercomputation*. Although no hypercomputer has yet been built (and perhaps never will be), it is instructive to consider what properties any such device should possess, and whether these requirements could ever be met. Aside from the potential benefits that would accrue from a positive outcome, the issues raised are sufficiently disruptive that they force us to re-evaluate existing computability theory. From a foundational viewpoint the questions driving hypercomputation theory remain the same as those addressed since the earliest days of computer science, viz. *what is computation?* and *what can be computed?* Early theoreticians developed models of computation that are independent of both their implementation and their physical location, but it has become clear in recent decades that these aspects of computation cannot always be neglected. In particular, the computational power of a distributed system can be expected to vary according to the spacetime geometry in which the machines on which it is running are located. The power of a computing system therefore depends on its physical environment and cannot be specified in absolute terms. Even Turing machines are capable of super-Turing behaviour, given the right environment.

Contents

Invited Papers

Decidability Problems for Self-induced Systems Generated
by a Substitution   3
   Timo Jolivet and Anne Siegel

Towards Formal Verification of Computations and Hypercomputations
in Relativistic Physics 17
   Mike Stannett

Regular Papers

A Connection Between Red-Green Turing Machines
and Watson-Crick T0L Systems 31
   Erzsébet Csuhaj-Varjú, Rudolf Freund, and György Vaszil

Tight Bounds for Cut-Operations on Deterministic Finite Automata 45
   Frank Drewes, Markus Holzer, Sebastian Jakobi,
   and Brink van der Merwe

Non-isometric Contextual Array Grammars with Regular Control
and Local Selectors 61
   Henning Fernau, Rudolf Freund, Rani Siromoney,
   and K.G. Subramanian

Universality of Graph-controlled Leftist Insertion-deletion Systems
with Two States 79
   Sergiu Ivanov and Sergey Verlan

Tinput-Driven Pushdown Automata 94
   Martin Kutrib, Andreas Malcher, and Matthias Wendlandt

Reversible Limited Automata 113
   Martin Kutrib and Matthias Wendlandt

An Intrinsically Universal Family of Causal Graph Dynamics 129
   Simon Martiel and Bruno Martin

The Simulation Powers and Limitations of Hierarchical
Self-Assembly Systems 149
   Jacob Hendricks, Matthew J. Patitz, and Trent A. Rogers

A Characterization of NP Within Interval-Valued Computing   164
   Benedek Nagy and Sándor Vályi

Universality in Infinite Petri Nets   180
   Dmitry A. Zaitsev

Author Index   199

Invited Papers

Decidability Problems for Self-induced Systems Generated by a Substitution

Timo Jolivet[1] and Anne Siegel[2,3]([✉])

[1] Université Paul Sabatier, Institut de Mathmatique de Toulouse, Toulouse, France
[2] CNRS, Université de Rennes 1, IRISA-UMR 6074, Rennes, France
anne.siegel@irisa.fr
[3] Inria Rennes - Bretagne Atlantique, Team Dyliss, Rennes, France

Abstract. In this talk we will survey several decidability and undecidability results on topological properties of self-affine or self-similar fractal tiles. Such tiles are obtained as fixed point of set equations governed by a graph. The study of their topological properties is known to be complex in general: we will illustrate this by undecidability results on tiles generated by multitape automata. In contrast, the class of self affine tiles called Rauzy fractals is particularly interesting. Such fractals provide geometrical representations of self-induced mathematical processes. They are associated to one-dimensional combinatorial substitutions (or iterated morphisms). They are somehow ubiquitous as self-replication processes appear naturally in several fields of mathematics. We will survey the main decidable topological properties of these specific Rauzy fractals and detail how the arithmetic properties of the substitution underlying the fractal construction make these properties decidable. We will end up this talk by discussing new questions arising in relation with continued fraction algorithm and fractal tiles generated by S-adic expansion systems.

The following survey is mainly inspired by three papers from the authors and their collaborators [32, 60, 84].

1 Substitutions Among Mathematics and Computer Science

A *substitution* (sometimes also called *iterated morphism*) is a combinatorial object which produces sequences by iteration. It is given by a replacement rule of the letters of a finite alphabet by nonempty, finite words over the same alphabet. Thus substitutions define an iteration process on a finite set in a natural way: as we shall detail it below, they appear in many fields of mathematics, theoretical physics and computer science whenever repetitive processes or replacement rules occur.

In *combinatorics on words*, substitutions have been used in order to exhibit examples of finite words or infinite sequences with very specific or unusual combinatorial properties. The most famous example is the Thue-Morse sequence,

© Springer International Publishing Switzerland 2015
J. Durand-Lose and B. Nagy (Eds.): MCU 2015, LNCS 9288, pp. 3–16, 2015.
DOI: 10.1007/978-3-319-23111-2_1

that is, the infinite fixed-point of the substitution $\sigma(1) = 12$, $\sigma(2) = 21$, which is the first example of an overlap-free infinite sequence, meaning that it contains no subword of the shape $1u1u1$, where $u \in \{1, 2\}^*$ [25,49,52,71,87,88]. The famous class of *Sturmian sequences*, including the famous Fibonacci substitution, is also strongly related to the composition of substitutions in relation with continued fraction expansion (see [49, Chapter 6]). Their combinatorial properties are particularly interesting, in terms of minimal complexity as well as in terms of representation of discrete lines. A sub-class of sturmian sequences with a quadratic irrational ratio are even proved to be a fixed point of a substitution [92]. More generally, the use of substitutions to describe discrete planes in $\mathbb{R}^3$ has proved to be very useful in discrete geometry to decide algorithmically whether a discrete patch is a part of a given discrete plane [12,28,48].

Since the time when they first appeared, substitutions have also been deeply related to *number theory*. For instance, in the field of diophantine approximation, substitutions produce transcendental numbers which can be approximated by cubic algebraic integers only in a very bad way [81]; the description of greedy expansions of reals in noninteger base [5,89] by means of substitutions also results in best approximation characterizations [53,71]. The Cobham Theorem [41] also constitutes a strong bridge between substitutions and number theory and allows one to derive deep transcendence properties: the real numbers with continued fraction expansions given by the Thue-Morse sequence, the Baum-Sweet sequence [22] or the Rudin-Shapiro sequence [82] are all transcendental, the proof being based on the "substitutive" structure of these sequences [1].

Substitutions also appear in *theoretical physics* in connection with quasicrystals, a class of crystals with forbidden symmetry [34,72,83]. Roughly, a solid is usually considered as a quasicrystal when it has an essentially discrete diffraction diagram. From a more combinatorial point of view, a quasicrystal is given by an aperiodic but repetitive structure that plays the role of the lattice in the theory of crystalline structures. Mathematically, we then speak of *Meyer sets* which are obtained by employing the cut-and-project scheme [76]. In the one-dimensional case, a well-studied family of Meyer sets is given by analogs of the integers in radix representations with respect to a non-integral base, in relation with sturmian sequences [51]. In higher-dimensional cases, however, the situation becomes much more difficult. The well-known Penrose tiling is a quasicrystal since it has essentially discrete diffraction diagram, but defining a wide class of examples of quasicrystals is an open question [10,62,63,80]. Good candidates for cut-and-project schemes (hence, quasicrystals) are given by discrete approximations of planes that are orthogonal to Pisot directions [11,23] and can be generated from one-dimensional substitutions with continued fraction algorithms [11,56]. Following this construction process, focusing on the periodic cases—an analog to the quadratic case within the sturmian family—it has been proved that substitutions provide relevant classes of examples for such Meyer sets in the multi-dimensional case, resulting in explicit examples of atomic structures (or point sets) with a relevant discrete diffraction diagram [30].

Another independent reason for the introduction of substitutions is related to *dynamical systems*, and more precisely to the field of *symbolic dynamics*, that is, the study of dynamical systems by coding their orbits as infinite sequences; to this matter, a complicated dynamics over a quite simple space is replaced by a simple dynamics (the shift mapping) over an intricate but combinatorial space made of infinite sequences. For the complete class of dynamical systems for which past and future are disjoint, the symbolic dynamical systems are particularly simple: they are described by a finite number of forbidden words, and they are called *shifts of finite type* [68]. A partition that induces a coding from a dynamical system onto such a shift of finite type is called a *Markov partition* [3]. The existence of Markov partitions is extremely useful in studying many dynamical properties (especially statistical ones); they are used for instance in analytic number theory of in dynamical systems [4, 36]. Explicit Markov partitions, however, are generally known only for hyperbolic automorphisms of the two-dimensional torus [3], and they have rectangular shapes. In higher dimensions, a slightly different behavior appears since several results attest that the contracting boundary of a member of a Markov partition cannot be smooth [37, 40]. In this setting, substitutions have been proved to be useful to construct explicit Markov partitions, based on generalized radix representations with a matrix as base (derived from the substitutions) [63, 65–67, 77] or referring to two-dimensional iteration processes (discrete planes construction processes discussed above) [14, 55].

In contrast to shifts of finite type we mention highly ordered self-similar systems with zero entropy, which can be defined as systems where the large-scale recurrence structure is similar to the small-scale recurrence structure, or more precisely, as systems which are topologically conjugate to their first return mapping on a particular subset. Importantly, such dynamical systems connected to all fields described above: they appear as a return mapping of the expanding flow onto the contracting manifold for hyperbolic toral automorphisms with a unique expanding direction [20] (in relation with dynamical systems and Markov partitions); they allow one to generate infinite sequences with interesting combinatorial features (in relation with combinatorics of words); they provide a construction of points sets for quasicristals (in relation with theoretical physics); they also allow one to describe non unique greedy expansions of reals in non integer basis. From their self-similar structure, their symbolic dynamical systems are naturally generated by substitutions [49, 78]. Therefore, substitutions appear to be ubiquitous to describe self-repetitive mathematical processes.

Paper Organization. In Sect. 2 we define *Rauzy fractals* and explain how they are used in the study of one-dimensional substitution dynamics. We highlight how the topological properties of these sets are used to study substitutions under several different viewpoints. In Sect. 3 we describe a general framework to study fractal objects, namely *graph-directed iterated function systems (GIFS)*. This class of objects includes Rauzy fractals. We see that many natural topological properties are undecidable in this framework. In Sect. 4 we give several examples of properties which are undecidable in the GIFS framework, but which are in decidable for the particular family of Rauzy fractals, and we briefly describe the main ideas behind some of the algorithms. In Sect. 5 we consider the more general problem of understanding the infinite family of systems generated by all the products of substitutions from a given finite set, and how we can extend the existing tools to study study such families.

2   The Geometry of One-Dimensional Substitutions

In the world of substitutions, geometrical objects appeared in 1982 in the work of Rauzy [79], to build a domain exchange in $\mathbb{R}^2$ that generalizes the theory of interval exchange transformations [61,91].

To each substitution one can associate an *incidence matrix* $\mathbf{M} = (m_{ij})$ in a natural way. Indeed, m_{ij} counts the occurrences of the letter i in $\sigma(j)$. To build a *Rauzy fractal*, we restrict to the case of *unit Pisot substitutions*, *i.e.*, substitutions whose incidence matrix is primitive and has a Pisot unit as a dominant eigenvalue.

An approach to build a Rauzy fractal is based on formal power series and projections of broken lines to hyperplanes and is inspired by Rauzy's seminal paper [79]. The principle is to consider a periodic point for the substitution, then to represent this sequence as a stair (also called "broken line") in $\mathbb{R}^n$, where n denotes the size of the alphabet on which the substitution acts. The next step is to project the vertices of the stair onto the contracting subspace of the incidence matrix. Since the projection is performed on a contracting stable space of the matrix, and the object that was projected is a periodic point of the substitution (and, hence, "contracted" by the incidence matrix) the closure of the projection is a compact set. A final step consists in drawing several colors with respect to the direction used in the stair to arrive on each vertex before the projection, and we get the Rauzy fractal.

The standard example is given by the so-called *Tribonacci substitution* defined as $\sigma(1) = 12$, $\sigma(2) = 13$, $\sigma(3) = 1$ which was first studied by Rauzy [79]. Projecting the "broken line" related to the unique fixed point of the Tribonacci substitution to the two-dimensional contracting plane yields a nice fractal picture, the so-called *classical Rauzy fractal* $\mathcal{T}$ which is depicted in Fig. 1 with its subtiles $\mathcal{T}(1)$ (largest subtile), $\mathcal{T}(2)$ (middle size subtile), $\mathcal{T}(3)$ (smallest subtile).

Since this compact set is obtained from the fixed point of the substitution, the self-replication properties of the fixed point have geometrical consequences: we represent the contracting space as the complex plane $\mathbb{C}$. Denote by α one of the two complex conjugate roots of the characteristic polynomial of the substitution matrix $X^3 - X^2 - X - 1$; one has $|\alpha| < 1$. With help of α, the Rauzy fractal can be written as graph directed iterated function system in the sense of [73] as

$$\begin{cases} \mathcal{T}(1) = \alpha(\mathcal{T}(1) \cup \mathcal{T}(2) \cup \mathcal{T}(3)), \\ \mathcal{T}(2) = \alpha\mathcal{T}(1) + 1, \\ \mathcal{T}(3) = \alpha\mathcal{T}(2) + 1. \end{cases}$$

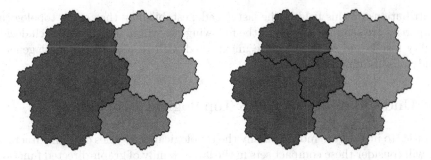

Fig. 1. The classical Rauzy fractal with its subtiles (left), and their self-affine decompositions (left).

Hence, each subtile is a finite union of translated contracted copies of subtiles. The contraction is given by the Galois conjugate α (of modulus <1), while the translations depend on the structure of the substitution.

The main reason for the frequent use of Rauzy fractals in the literature is that the iterative procedure to generate infinite words with the help of a substitution is often shifted to a geometric framework and reflects in self-similarity properties that can be studied. Then, the main questions to be investigated in each domain can be interpreted as questions related to the topology of the Rauzy fractal and its tiling properties.

In *number theory*, diophantine properties are induced by properties of a distance function to a specific broken line [53] related to the Rauzy fractal and the size of the largest ball contained in it. Finiteness properties of digit representations in *numeration systems* with non-integer base are related to the fact that **0** is an inner point of the Rauzy fractal [9]. More generally, the identification of those real numbers who has a periodic expansion in non-integer basis is strongly related to the study of the intersection of the fractal boundary with appropriate lines [2, 6]. Rauzy fractals also allow one to characterize purely periodic orbits of representations in numeration systems w.r.t. non-integer base, and yield certain generalizations of Galois' theorem [27, 30, 57]. In *discrete geometry*, there are numerous relations between generalized Rauzy fractals and discrete planes as studied for instance in [13]. The shape of pieces generating a discrete plane is tightly related to the shape of Rauzy fractals. On the *dynamical system* viewpoint, Rauzy fractals allow one to explicitly build the largest spectral factor induced by a substitutive dynamical system. Explicit Markov partitions for hyperbolic automorphisms of tori are constructed for instance in [55, 77], actually using Rauzy fractals. Importantly, in this setting, connectivity properties of Rauzy fractals are crucial to establish generator properties of the Markov partition [3]. In *tiling theory*, Rauzy fractals are used to represent the tiling flow and to prove that substitutive systems are expanding foliations of the space tiling [17].

For all these reasons, a thorough study of the topological properties of Rauzy fractals have appeared to be of great importance. There have been quite many contributions on this field in the last decade, establishing that many topological properties are semi-decidable. In the following, we will illustrate why such decidability results are somehow quite unexpected with respect to the most general study of self-affine tiles.

3   Undecidability of GIFS Topological Properties

In order to investigate more formally the topological properties of Rauzy fractals, we will consider these compact sets in the larger family of graph-directed function systems (GIFS).

One of the most common ways to define fractals is to use an iterated function system (IFS), defined by a finite collection of maps $f_1, \ldots, f_n : \mathbb{R}^d \to \mathbb{R}^d$ which are all contracting: there exists $0 \leq c < 1$ such that $\|f_i(x) - f_i(y)\| \leq c\|x - y\|$ for all $x, y \in \mathbb{R}^d$. The associated fractal, called the attractor of the IFS, is the unique nonempty compact set R such that

$$R = \bigcup_{i=1}^{n} f_i(R).$$

Such a set R always exists and is unique thanks to a famous result of Hutchinson [54], based on an application of Banach fixed-point theorem; see also [46] or [21]. For example, the classical Cantor set can be defined as the unique compact set $X \subseteq \mathbb{R}$ satisfying the set equation $X = \frac{1}{3}X \cup (\frac{1}{3}X + \frac{2}{3})$, and the Sierpiński triangle can defined as the unique compact set $X \subseteq \mathbb{R}^2$ satisfying $X = \frac{1}{2}X \cup (\frac{1}{2}X + (1/2, 0)) \cup (\frac{1}{2}X + (0, 1/2))$.

A natural generalization of IFS can be obtained by restricting which infinite sequences of maps $(f_{i_n})_{n \in \mathbb{N}}$ we are allowed to iterate. One of the simplest such restrictions is to require the set of allowed sequence $(i_n)_{n \in \mathbb{N}}$ to be the language of the infinite paths of a finite graph. Doing so we can give a new definition: a d-dimensional graph-directed iterated function system (GIFS) [73] is a directed graph in which each edge e is labelled by a contracting mapping $f_e : \mathbb{R}^d \to \mathbb{R}^d$.

It can be shown by a fixed point argument that given a GIFS $(G, \{f_e\}_{e \in E})$ there exists a unique collection of non-empty compact sets $\{R_q\}_{q \in \mathcal{Q}}$ such that

$$R_q = \bigcup_{q \in \mathcal{Q}} \bigcup_{e \in E_{q,r}} f_e(R_r),$$

where $\mathcal{Q}$ is the set of vertices of the directed graph defining the GIFS, and $E_{q,r}$ denote the set of edges from vertex q to vertex r. The sets R_i are called *GIFS attractors* or *solutions of the GIFS*. Note that the uniqueness statement does not hold for general sets, but only for non-empty compact sets [45].

In other words, the GIFS attractors are the solution of a set equation. Indeed, we are in presence of a collection of finitely many compact sets $\{R_1, \ldots, R_q\}$ such that each set R_i can be decomposed as a union of contracted copies of itself and the other sets R_j.

Many works are focused on the more specific family of *self-affine attractors*, in which the contractions f_i must be affine (of the form $M_i x + v_i$ where M_i is a $d \times d$ matrix and $v_i \in \mathbb{R}^d$), or the even more constrained family of self-similar attractors, in which the f_i must be similarities (of the form $ax + v_i$ where $a \in [0,1]$ and $v_i \in \mathbb{R}^d$).

Self-affine attractors are intensively studied, and many results are known about some particular families. For example the Hausdorff dimension of Bedford-McMullen carpets admits an exact simple formula [24,74], and similar results about the fractal dimension or the Lebesgue measure of some other classes exist [18,35,47,50,64]. Moreover, there is an "almost sure" formula for the packing and Hausdorff dimension in the self-similar case [44].

Despite all the positive results stated above, the notorious difficulty of self-affine sets suggests that there cannot exist any simple criteria to decide such properties in full generality. From a computer-theoretical point of view, this would correspond to undecidability results of the type: "there cannot be an algorithm that, given input an IFS specified by rational coefficients, determines if Property X holds for the IFS attractor", where "Property X" can be any IFS attractor property we are interested in. One could naturally expect a Rice-like theorem stating that every nontrivial property of GIFS attractor is undecidable, but such a statement does not hold. For example, the property of being equal to a singleton is nontrivial and decidable.

A first undecidability result has been established by Dube [42]: it is undecidable if the attractor of a rational 2-dimensional affine IFS intersects the diagonal $\{(x,x) : x \in [0,1]\}$.

In [60], the undecidability of some topological properties of self-affine graph-directed iterated function systems is established. More precisely, the authors prove that deciding whether the attractor of a 2-dimensional, 3-state affine GIFS has an empty interior is undecidable. In addition, deciding whether the intersection of two GIFS attractors $R_q \cap R_{q'}$ has empty interior is also undecidable.

To do so, the authors rely on the approach of Dube [42] and associate self-affine sets with computational devices called *multitape automata*, which are finite automata acting on several tapes, with an independent head reading each tape. Then they relate some properties of the automaton with topological properties of its associated attractor, and they obtain the undecidability of the latter by proving the undecidability of the former.

4 Decidability of Rauzy Fractals Properties

Equation 2 shows that Rauzy fractals also belong to the class of GIFS. In general, a Rauzy fractal associated with a substitution σ on the alphabet $\mathcal{A}$ satisfies the set equation:

$$\forall i \in A, \ \mathcal{T}(i) = \bigcup_{\substack{j \in A, \\ \sigma(j)=pis}} h(\mathcal{T}(j)) + \pi(p) \ . \tag{1}$$

Where h is a contraction map and π a mapping from $\mathcal{A}^*$ to the Euclidean space where the fractal lives. The graph with nodes in $\mathcal{A}$ and with edges described by the relation $\sigma(j) = pis$ is the so-called *prefix-suffix graph*. It describes the way images of letters under σ can be decomposed [38, 39]. The mappings in the GIFS are contracting, thus the nonempty compact sets $T(i)$ satisfying Eq. 2 (and more generally Eq. 1) are uniquely determined [73]. They always satisfy the following properties:

- The direction of the expanding eigenvector of the incidence matrix is irrational [39].
- T as well as $T(i)$ is compact [73].
- T as well as $T(i)$ is the closure of its interior and has a non-zero measure [32, 86].
- The subtiles $T(i)$ induce a self-replicating multiple tiling of the contracting plane [15].

Similarly to the families studied in [44], the Hausdorff dimension of the boundary of a planar Rauzy fractal can be computed if its corresponding Pisot eigenvalue has complex conjugates, because the associated GIFS is then self-similar [84]. However, no formula is known if the two conjugates of the Pisot eigenvalue are real, because then their norms are not equal, so the GIFS is self-affine but not self-similar. This is in agreement with the known difficulty of the study of self-affine sets.

On the contrary, the study of inner points and intersection between tiles is at the opposite of multitape automata: indeed, there are several algorithms to decide if the tiles of the Rauzy fractal of a unimodular Pisot substitution σ do not overlap, that is, if they intersect on a set of Lebesgue measure zero. In the case of Rauzy fractals, this is equivalent to having intersection with empty interior. It follows that the undecidable property that is stated for 2-dimensional, 3-state affine GIFS [60] is actually decidable for the case of Rauzy fractals (see the review in [32]). This is not contradictory since the family of Rauzy fractals GIFS is disjoint from the family of the GIFS associated with multitape automata for which undecidability results are proved. Indeed, negative powers of integers cannot be the expansion factors of a Rauzy fractal GIFS, which, in opposition, is always the case for multitape automata GIFS.

More generally, it has been proved that many topological properties or Rauzy fractals are actually decidable.

- Checking whether the origin is an inner point of T is decidable [84].
- Checking whether the Rauzy fractal generates self-similar tiling of the plane is decidable [7, 8, 17, 43, 58, 69, 70, 85].
- The box-counting dimension of the fractal boundary of the Rauzy fractal and its subtiles is computable. In the self-similar case, this allows computing the Haussorf-dimension [84, 90].
- Checking the connectivity of T and $T(i)$ is semi-decidable.
- Verifying that $T(i)$ is homeomorphic to a closed disk is semi-decidable.
- The non-triviality of the fundamental group of T is also semi-decidable, as well as the property of uncountability and being not free.

The underlying idea in all criteria is to match the structure of the graph directed iterated function system that defines the central tile with its tiling properties. All criteria make use and are expressed in terms of graphs. The graphs we are using to formulate and prove such results contain the structure of intersections of two or more tiles in the (multiple) tilings induced by the Rauzy fractal $\mathcal{T}$ and its subtiles $\mathcal{T}(i)$ $(1 \leq i \leq n)$. If the subtiles induce a tiling, they provide a description of the boundaries of the subtiles $\mathcal{T}(i)$ $(1 \leq i \leq n)$ and even permit to draw these boundaries in an easy way. Other graphs encode the connectivity of the Rauzy fractal, its subtiles as well as of certain pieces of their boundary.

5   Extending the Framework of Rauzy Fractals

As detailed below, topological properties of Rauzy fractals are now well understood and can be checked for each single Pisot unit substitution. Nonetheless, these decidability results rely on the construction of graphs which are deeply dependent on the combinatorics of the substitution and the algebraic properties of its incidence matrix. This raises a strong issue when one wishes to address general results about families of substitutions. Then, the decidability issue becomes: *Is there an algorithm that, given a finite family of substitutions, determines whether Property X holds for the Rauzy fractal of every finite product of substitutions in the input family of substitutions?* This is for example how Arnoux-Rauzy substitutions are constructed [16]. More generally, two-dimensional continued fraction algorithms (Brun and Jacobi-Perron continued fraction algorithms) defined as piecewise fractional maps produce product families of three-letter substitutions which seems to have relevant invariant topological properties [26].

To address such decidability questions about product families of substitutions, new frameworks need to be developed. Two main trends are studied nowadays. First, relying on the mathematical study of proximality and homoclinic return points in tiling flows, Barge proved the product family of β-substitutions, Brun substitutions and Jacobi-Perron substitutions all generate aperiodic tilings, meaning that their boundary can be approximated as the Haussdorf limit of polygonal transformations [19]. Second, the construction and study of local two-dimensional substitution rules to generate Rauzy fractals allow to study generic topological properties: for instance, these techniques allow one to solve the decidability problem of connectivity for Rauzy fractals associated to the product families of Arnoux-Rauzy, Brun and Jacobi-Perron substitutions (which is decidable), and simple-connectivity (which in non-decidable) [27,59]. As an application of this result, it becomes possible to elucidate which two-dimensional toral translation can be represented by a symbolic substitutive system, an extension of the representation of one-dimensional toral translation by sturmian sequences. Such a framework has also been successfully applied to a problem in discrete geometry about the critical thickness at which an arithmetic discrete plane is 2-connected [29].

Another raising issue is the study of fractal tiles generated by substitution-like processes although they do not exactly fit with the framework of Rauzy fractals or GIFS. In this trend, we mention tiles generated by non unit numbers [75] or SRS-numerations systems [31], that have intricate properties, and tiles generated by infinite compositions of substitution (S-adic) framework [33]. In both cases, all remains to be done in terms of decidability problems. Indeed, in most cases, the basis algebraic properties of the underlying systems (i.e., rational independency of translation vectors in the GIFS), or the basic topological properties (i.e. the fractal has a non empty interior, 0 is a inner point) cannot be established in general. Such properties will deserve algorithmic studies and pave the way to very exciting novel issues to address.

References

1. Adamczewski, B., Bugeaud, Y., Davison, L.: Continued fractions and transcendental numbers. Ann. Inst. Fourier (Grenoble) **56**(7), 2093–2113 (2006). (Numération, pavages, substitutions)
2. Adamczewski, B., Frougny, C., Siegel, A., Steiner, W.: Rational numbers with purely periodic β-expansion. Bull. Lond. Math. Soc. **42**(3), 538–552 (2010)
3. Adler, R.L.: Symbolic dynamics and Markov partitions. Bull. Amer. Math. Soc. (N.S.) **35**(1), 1–56 (1998)
4. Adler, R.L., Weiss, B.: Similarity of automorphisms of the torus. Memoirs of the American Mathematical Society, No. 98. American Mathematical Society, Providence, R.I (1970)
5. Akiyama, S.: Pisot numbers and greedy algorithm. In: Number theory (Eger, 1996), pp. 9–21. de Gruyter, Berlin (1998)
6. Akiyama, S., Barat, G., Berthé, V., Siegel, A.: Boundary of central tiles associated with Pisot beta-numeration and purely periodic expansions. Monatsh. Math. **155**(3–4), 377–419 (2008)
7. Akiyama, S., Lee, J.-Y.: Algorithm for determining pure pointedness of self-affine tilings. Adv. Math. **226**(4), 2855–2883 (2011)
8. Akiyama, S., Lee, J.-Y.: Overlap coincidence to strong coincidence in substitution tiling dynamics. Eur. J. Combin. **39**, 233–243 (2014)
9. Akiyama, S., Scheicher, K.: Intersecting two-dimensional fractals with lines. Acta Sci. Math. (Szeged) **71**(3–4), 555–580 (2005)
10. Anderson, J., Putnam, I.: Topological invariants for substitution tilings and their associated C^*-algebras. Ergodic Theory Dyn. Syst. **18**, 509–537 (1998)
11. Arnoux, P., Berthé, V., Ei, H., Ito, S.: Tilings, quasicrystals, discrete planes, generalized substitutions, and multidimensional continued fractions. In: Discrete models: combinatorics, computation, and geometry (Paris, 2001). Discrete Math. Theor. Comput. Sci. Proc., AA, pages 059–078 (electronic). Maison Inform. Math. Discrèt. (MIMD), Paris (2001)
12. Arnoux, P., Berthé, V., Fernique, T., Jamet, D.: Functional stepped surfaces, flips, and generalized substitutions. Theor. Comput. Sci. **380**(3), 251–265 (2007)
13. Arnoux, P., Berthé, V., Ito, S.: Discrete planes, $\mathbb{Z}^2$-actions, Jacobi-Perron algorithm and substitutions. Ann. Inst. Fourier **52**(2), 305–349 (2002)
14. Arnoux, P., Furukado, M., Harriss, E., Ito, S.: Algebraic numbers, group automorphisms and substitution rules on the plane. Trans. Amer. Math. Soc. (2009, to appear)

15. Arnoux, P., Ito, S.: Pisot substitutions and Rauzy fractals. Bull. Belg. Math. Soc. Simon Stevin **8**(2), 181–207 (2001)
16. Arnoux, P., Rauzy, G.: Représentation géométrique de suites de complexité $2n+1$. Bull. Soc. Math. Fr. **119**(2), 199–215 (1991)
17. Baker, V., Barge, M., Kwapisz, J.: Geometric realization and coincidence for reducible non-unimodular Pisot tiling spaces with an application to beta-shifts. Ann. Inst. Fourier **56**(7), 2213–2248 (2006)
18. Barański, K.: Hausdorff dimension of the limit sets of some planar geometric constructions. Adv. Math. **210**(1), 215–245 (2007)
19. Barge, M.: Pure discrete spectrum for a class of one-dimensional substitution tiling systems (2014)
20. Barge, M., Kwapisz, J.: Geometric theory of unimodular Pisot substitutions. Amer. J. Math. **128**(5), 1219–1282 (2006)
21. Barnsley, M.F.: Fractals Everywhere, 2nd edn. Academic Press Professional, Boston (1993)
22. Baum, L.E., Sweet, M.M.: Continued fractions of algebraic power series in characteristic 2. Ann. Math. (2) **103**(3), 593–610 (1976)
23. Béal, M.-P., Perrin, D.: Symbolic dynamics and finite automata. In: Rozenberg, G., Salomaa, A. (eds.) Handbook of Formal Languages. Linear Modeling: Background and Application, pp. 463–506. Springer, Heidelberg (1997)
24. Bedford, T.: Crinkly curves, Markov partitions and box dimensions in self-similar sets. Ph.D. thesis, University of Warwick (1984)
25. Berstel, J., Perrin, D.: The origins of combinatorics on words. Eur. J. Comb. **28**(3), 996–1022 (2007)
26. Berthé, V.: Multidimensional Euclidean algorithms, numeration and substitutions. Integers, 11B: Paper No. A2, 34 (2011)
27. Berthé, V., Bourdon, J., Jolivet, T., Siegel, A.: A combinatorial approach to products of Pisot substitutions. Ergodic Theory and Dynamical Systems (2015, to appear)
28. Berthé, V., Fernique, T.: Brun expansions of stepped surfaces. Discrete Math. **311**(7), 521–543 (2011)
29. Berthé, V., Jamet, D., Jolivet, T., Provençal, X.: Critical connectedness of thin arithmetical discrete planes. In: Gonzalez-Diaz, R., Jimenez, M.-J., Medrano, B. (eds.) DGCI 2013. LNCS, vol. 7749, pp. 107–118. Springer, Heidelberg (2013)
30. Berthé, V., Siegel, A.: Tilings associated with beta-numeration and substitutions. Integers **5**, 46 (2005)
31. Berthé, V., Siegel, A., Steiner, W., Surer, P., Thuswaldner, J.M.: Fractal tiles associated with shift radix systems. Adv. Math. **226**(1), 139–175 (2011)
32. Berthé, V., Siegel, A., Thuswaldner, J.M.: Substitutions, Rauzy fractals, and tilings. In: Combinatorics, Automata and Number Theory. Encyclopedia of Mathematics and its Applications, vol. 135. Cambridge University Press (2010)
33. Berthé, V., Steiner, W., Thuswaldner, J.: Geometry, dynamics, and arithmetic of S-adic shifts. Article submitted for publication (2014)
34. Bombieri, E., Taylor, J.E.: Which distributions of matter diffract? An initial investigation. J. Phys. **47**(7, Suppl. Colloq. C3), C3-19–C3-28 (1986). (International workshop on aperiodic crystals, Les Houches (1986))
35. Bondarenko, I.V., Kravchenko, R.V.: On Lebesgue measure of integral self-affine sets. Discrete Comput. Geom. **46**(2), 389–393 (2011)
36. Bowen, R.: Equilibrium States and the Ergodic Theory of Anosov Diffeomorphisms, vol. 470. Springer, Heidelberg (1978)

37. Bowen, R.: Markov partitions are not smooth. Proc. Amer. Math. Soc. **71**(1), 130–132 (1978)
38. Canterini, V., Siegel, A.: Automate des préfixes-suffixes associé à une substitution primitive. J. Théor. Nombres Bordeaux **13**(2), 353–369 (2001)
39. Canterini, V., Siegel, A.: Geometric representation of substitutions of Pisot type. Trans. Am. Math. Soc. **353**(12), 5121–5144 (2001)
40. Cawley, E.: Smooth Markov partitions and toral automorphisms. Ergodic Theory Dyn. Syst. **11**(4), 633–651 (1991)
41. Cobham, A.: Uniform tag sequences. Math. Syst. Theory **6**, 164–192 (1972)
42. Dube, S.: Undecidable problems in fractal geometry. Complex Syst. **7**(6), 423–444 (1993)
43. Ei, H., Ito, S., Rao, H.: Atomic surfaces, tilings and coincidences II. reducible case. Ann. Inst. Fourier **56**, 2285–2313 (2006)
44. Falconer, K.: The Hausdorff dimension of self-affine fractals. Math. Proc. Camb. Philos. Soc. **103**(2), 339–350 (1988)
45. Falconer, K.: Techniques in Fractal Geometry. Wiley, Chichester (1997)
46. Falconer, K.: Fractal Geometry, Mathematical Foundations and Applications, 2nd edn. Wiley, Hoboken (2003)
47. Feng, D.-J., Wang, Y.: A class of self-affine sets and self-affine measures. J. Fourier Anal. Appl. **11**(1), 107–124 (2005)
48. Fernique, T.: Generation and Recognition of digital planes using multi-dimensional continued fractions. In: Coeurjolly, D., Sivignon, I., Tougne, L., Dupont, F. (eds.) DGCI 2008. LNCS, vol. 4992, pp. 33–44. Springer, Heidelberg (2008)
49. Fogg, N.P., Berthé, V., Ferenczi, S., Mauduit, C., Siegel, A. (eds.): Substitutions in Dynamics, Arithmetics and Combinatorics, vol. 1794. Springer, Heidelberg (2002)
50. Fraser, J.M.: On the packing dimension of box-like self-affine sets in the plane. Nonlinearity **25**(7), 2075–2092 (2012)
51. Gazeau, J.-P., Verger-Gaugry, J.-L.: Geometric study of the beta-integers for a Perron number and mathematical quasicrystals. J. Théor. Nombres Bordeaux **16**(1), 125–149 (2004)
52. Hedlund, G.A.: Remarks on the work of Axel Thue on sequences. Nordisk Mat. Tidskr. **15**, 148–150 (1967)
53. Hubert, P., Messaoudi, A.: Best simultaneous Diophantine approximations of Pisot numbers and Rauzy fractals. Acta Arith. **124**(1), 1–15 (2006)
54. Hutchinson, J.E.: Fractals and self-similarity. Indiana Univ. Math. J. **30**(5), 713–747 (1981)
55. Ito, S., Ohtsuki, M.: Modified Jacobi-Perron algorithm and generating Markov partitions for special hyperbolic toral automorphisms. Tokyo J. Math. **16**(2), 441–472 (1993)
56. Ito, S., Ohtsuki, M.: Parallelogram tilings and Jacobi-Perron algorithm. Tokyo J. Math. **17**(1), 33–58 (1994)
57. Ito, S., Rao, H.: Purely periodic β-expansion with Pisot base. Proc. Am. Math. Soc. **133**, 953–964 (2005)
58. Ito, S., Rao, H.: Atomic surfaces, tilings and coincidences I. Irreducible case. Isr. J. Math. **153**, 129–155 (2006)
59. Jolivet, T., Kari, J.: Consistency of multidimensional combinatorial substitutions. In: Hirsch, E.A., Karhumäki, J., Lepistö, A., Prilutskii, M. (eds.) CSR 2012. LNCS, vol. 7353, pp. 205–216. Springer, Heidelberg (2012)
60. Jolivet, T., Kari, J.: Undecidable properties of self-affine sets and multi-tape automata. In: Csuhaj-Varjú, E., Dietzfelbinger, M., Ésik, Z. (eds.) MFCS 2014, Part I. LNCS, vol. 8634, pp. 352–364. Springer, Heidelberg (2014)

61. Keane, M.: Interval exchange transformations. Math. Z. **141**, 25–31 (1975)
62. Kellendonk, J., Putnam, I.: Tilings, C^*-algebras, and K-theory. In: Baake, M., Moody, R.V. (eds.) Directions in Mathematical Quasicrystals. AMS CRM Monogr. Ser., vol. 13, pp. 177–206, Providence, RI (2000)
63. Kenyon, R., Vershik, A.: Arithmetic construction of sofic partitions of hyperbolic toral automorphisms. Ergodic Theory Dyn. Syst. **18**(2), 357–372 (1998)
64. Lalley, S.P., Gatzouras, D.: Hausdorff and box dimensions of certain self-affine fractals. Indiana Univ. Math. J. **41**(2), 533–568 (1992)
65. Le Borgne, S.: Un codage sofique des automorphismes hyperboliques du tore. In: Séminaires de Probabilités de Rennes. Publ. Inst. Rech. Math. Rennes, vol. 1995, p. 35. University of Rennes 1, Rennes (1995)
66. Le Borgne, S.: Un codage sofique des automorphismes hyperboliques du tore. C.R. Acad. Sci. Paris Sér. I Math. **323**(10), 1123–1128 (1996)
67. Le Borgne, S.: Un codage sofique des automorphismes hyperboliques du tore. Bol. Soc. Brasil. Mat. (N.S) **30**(1), 61–93 (1999)
68. Lind, D., Marcus, B.: An Introduction to Symbolic Dynamics and Coding. Cambridge University Press, Cambridge (1995)
69. Livshits, A.N.: On the spectra of adic transformations of markov compacta. Russ. Math. Surv. **42**, 222–223 (1987)
70. Livshits, A.N.: Some examples of adic transformations and automorphisms of substitutions. Sel. Math. Sov. **11**(1), 83–104 (1998). Selected translations
71. Lothaire, M.: Applied Combinatorics on Words. Encyclopedia of Mathematics and its Applications, vol. 105. Cambridge University Press, Cambridge (2005)
72. Luck, J.M., Godrèche, C., Janner, A., Janssen, T.: The nature of the atomic surfaces of quasiperiodic self-similar structures. J. Phys. A **26**(8), 1951–1999 (1993)
73. Mauldin, R.D., Williams, S.C.: Hausdorff dimension in graph directed constructions. Trans. Am. Math. Soc. **309**(2), 811–829 (1988)
74. McMullen, C.: The Hausdorff dimension of general Sierpiński carpets. Nagoya Math. J. **96**, 1–9 (1984)
75. Minervino, M., Steiner, W.: Tilings for Pisot beta numeration. Indag. Math. (N.S.) **25**(4), 745–773 (2014)
76. Moody, R.V.: Model sets: a survey. In: Axel, F., Dénoyer, F., Gazeau, J.-P. (eds.) From Quasicrystals to More Complex Systems. Centre de Physique des Houches, vol. 13, pp. 145–166. Springer, Heidelberg (2000)
77. Praggastis, B.: Numeration systems and Markov partitions from self-similar tilings. Trans. Am. Math. Soc. **351**(8), 3315–3349 (1999)
78. Queffélec, M.: Substitution Dynamical Systems-Spectral Analysis, vol. 1294. Springer, Heidelberg (1987)
79. Rauzy, G.: Nombres algébriques et substitutions. Bull. Soc. Math. Fr. **110**(2), 147–178 (1982)
80. Robinson, Jr., E.A.: Symbolic dynamics and tilings of $\mathbb{R}^d$. In: Symbolic dynamics and its applications. Proc. Sympos. Appl. Math., Amer. Math. Soc., vol. 60, pp. 81–119, Providence, RI (2004)
81. Roy, D.: Approximation to real numbers by cubic algebraic integers. II. Ann. Math. (2) **158**(3), 1081–1087 (2003)
82. Rudin, W.: Some theorems on Fourier coefficients. Proc. Am. Math. Soc. **10**, 855–859 (1959)
83. Senechal, M.: What is... a quasicrystal? Not. Am. Math. Soc. **53**(8), 886–887 (2006)
84. Siegel, A., Thuswaldner, J.M.: Topological properties of Rauzy fractals. Mém. Soc. Math. Fr. (N.S.) **118**, 140 (2009)

85. Sirvent, V.F., Solomyak, B.: Pure discrete spectrum for one-dimensional substitution systems of Pisot type. Canad. Math. Bull. **45**(4), 697–710 (2002). (Dedicated to Robert V. Moody)

86. Sirvent, V.F., Wang, Y.: Self-affine tiling via substitution dynamical systems and Rauzy fractals. Pac. J. Math. **206**(2), 465–485 (2002)

87. Thue, A.: Über unendliche Zeichenreihen. Norske Vid. Selsk. Skr. Mat. Nat. Kl. **37**(7), 1–22 (1906)

88. Thue, A.: Über die gegenseitige Lage gleicher Teile gewisser Zeichenreihen. Norske Vid. Selsk. Skr. Mat. Nat. Kl. **43**(1), 1–67 (1912)

89. Thurston, W.P.: Groups, tilings and finite state automata. Lectures notes distributed in conjunction with the Colloquium Series, in AMS Colloquium lectures (1989)

90. Thuswaldner, J.M.: Unimodular Pisot substitutions and their associated tiles. Journal de Theorie des Nombres de Bordeaux **18**(2), 487–536 (2006)

91. Veech, W.A.: Interval exchange transformations. J. Anal. Math. **33**, 222–272 (1978)

92. Yasutomi, S.-I.: On Sturmian sequences which are invariant under some substitutions. In: Number theory and its Applications (Kyoto, 1997). Dev. Math., vol. 2, pp. 347–373. Kluwer Academic Publishers, Dordrecht (1999)

Towards Formal Verification of Computations and Hypercomputations in Relativistic Physics

Mike Stannett[✉]

Department of Computer Science, University of Sheffield, Regent Court,
211 Portobello, Sheffield S1 4DP, Sheffield, UK
m.stannett@sheffield.ac.uk

Abstract. It is now more than 15 years since Copeland and Proudfoot introduced the term *hypercomputation*. Although no hypercomputer has yet been built (and perhaps never will be), it is instructive to consider what properties any such device should possess, and whether these requirements could ever be met. Aside from the potential benefits that would accrue from a positive outcome, the issues raised are sufficiently disruptive that they force us to re-evaluate existing computability theory. From a foundational viewpoint the questions driving hypercomputation theory remain the same as those addressed since the earliest days of computer science, viz. *what is computation?* and *what can be computed?* Early theoreticians developed models of computation that are independent of both their implementation and their physical location, but it has become clear in recent decades that these aspects of computation cannot always be neglected. In particular, the computational power of a distributed system can be expected to vary according to the spacetime geometry in which the machines on which it is running are located. The power of a computing system therefore depends on its physical environment and cannot be specified in absolute terms. Even Turing machines are capable of super-Turing behaviour, given the right environment.

1 Introduction

The term hypercomputation refers to the study of physical or abstract systems which are potentially capable of behaviours which cannot be simulated by recursive means. The term was introduced by Copeland and Proudfoot ([2]) as a more accurate replacement for the term 'super-Turing' used by Stannett ([13–15]) and Siegelmann ([12]) to describe certain types of putative hypercomputational system. Although no hypercomputer has yet been built (and perhaps never will be), it is instructive to consider what properties any such device should possess, and whether these requirements could ever be met.

Computers are physical devices whose possible behaviours are constrained and described by physical laws. The answers to the questions *what can be computed?* and *what can be computed quickly?* therefore depend on ones theory of physics and the properties of physical materials. Moreover, because physical devices exist in space and time, their computational power can depend both on

© Springer International Publishing Switzerland 2015
J. Durand-Lose and B. Nagy (Eds.): MCU 2015, LNCS 9288, pp. 17–27, 2015.
DOI: 10.1007/978-3-319-23111-2_2

when and where they are located. In particular, spacetime structures can boost the power of computational systems, but can also constrain and reduce their power. Similarly, an algorithm's run-time complexity is not an absolute property but depends on the spacetime trajectory being followed by the machine(s) on which it is running.

1.1 Geometrical Boosting of Computational Power

A well-known strategy for boosting computational power is to exploit the properties of *Malament-Hogarth* (M-H) spacetimes [3]. These are spacetimes containing a point p and a future-pointing semi-infinite worldline w not passing through p, such that every point x of w can be joined to p by a future-pointing timelike path which has finite proper length (Fig. 1). We refer to the pair (w, p) as an *M-H structure* in what follows.

The following lemma shows that all Σ_1^0 and Π_1^0 sets become decidable in M-H spacetime using just two Turing machines, provided they can communicate at least once.

Lemma 1. *Let S be any set in Σ_1^0 or Π_1^0. Then S can be decided in M-H spacetime by a system comprising two computers capable of communicating once.*

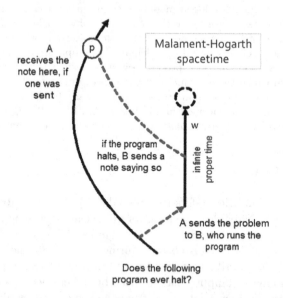

Fig. 1. Temporal structure of a hypercomputation using an M-H structure (w, p). In this example, we solve the Halting Problem in constant time using two communicating Turing machines. Machine A sends the program to machine B, and then travels to the M-H event p. Machine B, moving along w, runs the program and if it ever halts it sends a message to p saying so. On reaching p, A looks for the message. It is present at p if and only if the program halted.

Sender	Receiver
| ```
y = 0;
while (travelling along w) {
    while ( R(n,y) ) { y = y+1; }
    transmit (result = false) to p;
    halt;
}
``` | ```
result = true;
travel to p;
wait 1 second;
return result;
``` |

Fig. 2. The programs *Sender* (running on T_S, which is capable of sending at most one message to *Receiver*) and *Receiver* (running on T_R, which is capable of receiving and acting upon at most one message from *Sender*) co-operate to decide the undecidable set S in the context of an M-H structure (w, p). The two machines are initially co-located at some point on the worldline w. The 1-second delay is to avoid ambiguity as to whether *Receiver* returns *result* before or after executing *Sender*'s assignment instruction at p.

Proof. We show that any S in Π_1^0 can be decided in M-H spacetime (the Σ_1^0 case follows by complementarity). Since S is in Π_1^0 we can write $S = \{x \mid \forall y. R(x, y)\}$, where R is recursive. To decide whether $n \in S$, we run the programs *Sender* and *Receiver* shown in Fig. 2.

Suppose $n \notin S$, i.e. $\neg\forall y. R(n, y)$. Then there exists some y for which the test $R(n, y)$ fails. Let $y_{\min}$ be the smallest such y. Then

- The machine T_S travels along w, a trajectory which allows it infinite execution time (since it has infinite proper length). Consequently, *Sender* eventually encounters and fails the test $R(n, y_{\min})$, transmits the instruction "*result =* **false**" to p (along a trajectory of finite proper-length), and terminates.
- *Receiver* sets *result* to **true**, then travels to p where it encounters and executes the instruction sent there by *Sender* setting *result* to **false**. It waits one second and then returns the value of *result*, i.e. **false**.

Now suppose conversely that $n \in S$. Then

- *Sender* never exits the loop testing $R(n, y)$ and never issues the instruction setting *result* to **false**. It runs forever without terminating (its trajectory along w ensures that this is possible).
- *Receiver* sets *result* to **true** and travels to p. After waiting one second it returns the unchanged initial value of *result*, i.e. **true**.

In either case, the system eventually returns a value, and the value returned correctly reports whether or not $n \in S$. □

Lemma 1 shows that spacetime geometries can boost computational power, and that this does not require the introduction of 'unphysical' constructs like infinite precision observations or new types of machine. The machines used for this hypercomputation are simply Turing machines – indeed, *Receiver* is so simple that T_R could arguably be replaced by an essentially trivial 2-state automaton with no loss of power to the system as a whole. Notice, however, that a single machine acting alone cannot exploit the boosting effect of M-H structures, because this relies on splitting the system into two parts, one of which can run forever in a period of time that appears finite to the other. Notice also that spacetime geometries can be considerably more complicated than those considered here, and that structures can be envisaged which allow decidability at all levels of the arithmetic hierarchy [6] and beyond [18].

1.2   Geometrical Reduction of Computational Power

Spacetime geometry can also constrain and reduce computational power. For example, consider a computer traversing a closed timelike curve (CTC) or 'time loop'. Suppose the computer's clock shows that each circuit of the CTC is long enough for it to execute N instructions. Since the computer and all of its components return to their initial spacetime locations (and hence their initial machine states) after every N instructions, the number of steps executable by a CTC-traversing Turing machine is necessarily bounded, and all CTC-located programs must be reversible [16]. Indeed, it is only possible to run a fully controlled program if the temporal length of the CTC is an exact integer multiple of the program's runtime, since it will not otherwise return to its initial state on completion of each circuit.

1.3   Geometrical Effects on Computational Complexity

The possibility of M-H spacetimes also has implications for computational complexity. A simple adaptation of the distributed computation outlined in Lemma 1 allows the result produced by any program to be obtained within a fixed time period, viz. precisely one second longer than it takes *Receiver* to reach p. In M-H spacetimes, all programs have constant run-time complexity. (Similarly, CTCs can be use to transmit results 'into the past', thereby allowing program results to be obtained more quickly than would otherwise be the case.)

Notice, however, that this requires us to refine our notions of complexity slightly. The program itself may have arbitrarily large complexity, but it is running on the machine *Sender* which is not responsible for reporting the program output. Instead, this is reported by *Receiver* in constant time. In relativistic settings, it is essential to identify carefully which components in a distributed system are deemed responsible for generating the final system output.

2   Modelling Relativity Theory in Isabelle/HOL

Since a spacetime might potentially contain a combination of 'normal' regions, M-H structures and CTCs, the question "what can be computed" has no absolute answer but depends on local and global geometric properties, the number of machines available, their relative spacetime trajectories during computation, and the availability of suitable communication channels. This is a question we would like to investigate in more detail, but we are hampered by the informal yet detailed nature of many proofs in relativity theory (and physics in general). The issue is particularly relevant because the black hole observed at the centre of our own galaxy Milky Way is potentially of the right type to be a habitat for M-H structures [4], and while such structures are obviously beyond our current technological capabilities to exploit, the mere possibility of their existence is enough to warrant a re-evaluation of the extent to which abstract computability and complexity theory give an accurate account of what is actually possible in the physical universe.

In 2012 we joined forces with researchers at the Rényi Institute of Mathematics in Budapest, who have spent many years developing versions of relativity theory expressed in first order logic – our goal is to express the Hungarian theories in Isabelle/HOL [9] so as to allow machine-assisted investigation of various key hypotheses concerning the possibilities for computation and hypercomputation in relativistic physics [17]. In this section we briefly describe the Hungarian approach, and show how it can be translated with relative ease into machine-readable form.

2.1 First-Order Relativity Theory

The approach adopted by Andréka, Németi and the Hungarian team is to formulate a collection of related relativity theories in first-order logic (FOL), using axioms that are as simple and transparent as possible [1]. Our own starting point is the translation of the Hungarian axioms and theorems into machine-readable format suitable for use with the Isabelle/HOL proof assistant [9].

For example, special relativity is represented as a theory SPECREL based on just four physical axioms:

- AxPh (Photon Axiom)
  Each inertial observer considers the speed of light to be positive, and the same in every spatial direction. Moreover, photons can be emitted in or arrive from any spatial direction.
- AxEv (Event Axiom)
  All observers inhabit the same universe, i.e. they consider the same events to take place (though possibly at different locations or times).
- AxSelf (Self Axiom)
  Inertial observers consider themselves to be stationary.
- AxSym (Symmetry Axiom)
  Whenever observers consider two events to be simultaneous, they agree as to the spatial distance between those two events – this allows observers to calibrate their rulers relative to one another.

The underlying theory has two basic sorts: *quantities* and *bodies*. Quantities are used to express distances and times, and are assumed to satisfy the axioms of a field. Bodies in SPECREL include *inertial observers* and *photons*, which are identified by predicates, e.g. $IObs(b)$ is *true* if and only if body b is an inertial

```
class Lines = Quantities + Vectors + Points
begin

    . . .


fun space2 :: "('a Point) ⇒ ('a Point) ⇒ 'a" where
  "space2 u v
     = (xval u - xval v)*(xval u - xval v)
     + (yval u - yval v)*(yval u - yval v)
     + (zval u - zval v)*(zval u - zval v)"

fun time2 :: "('a Point) ⇒ ('a Point) ⇒ 'a" where
  "time2 u v = (tval u - tval v)*(tval u - tval v)"
```

Fig. 3. Spatial and temporal distances are defined as properties of lines, and are used to calculate the speeds needed to move from one spacetime location to another. The class `Lines` is one of several classes bundled together to form the background context class `SpaceTime` which defines the geometrical structures needed to describe spacetime. These include quantities, vectors, points, cones, straight lines and planes.

observer, and likewise $Ph(b)$ indicates whether b is a photon. Central to all of the Hungarian versions of first-order relativity theory is the *worldview relation*, W, where $W(m, b, x)$ means that observer m considers body b to be present at location x.

These constructs are generally sufficient to allow the axioms to be specified. For example, we can use the field axioms to define functions $space^2$ and $time^2$ giving the (squared) spatial and temporal distances between two spacetime events (Fig. 3). Recalling that $IOb(m)$ means "m is an inertial observer", these in turn let us write AxPh as

$$IOb(m) \rightarrow (\exists v.((v > 0) \wedge (\forall xy.($$
$$(\exists p.(Ph(p) \wedge W(m, p, x) \wedge W(m, p, y)))$$
$$\leftrightarrow (space^2\ xy = (v * v) * (time^2\ xy))))))$$

In words: each inertial observer is associated with a positive speed v with the property that whenever any photon is considered by m to pass through two spacetime locations x and y, the (squared) speed associated with the straight line joining these points is v^2.

The translation into Isabelle/HOL format is now straightforward, viz.

```
class AxPh = WorldView +
assumes
  AxPh:"IOb(m)
        ⟹ (∃v. ( (v > (0::'a)) ∧ ( ∀x y . (
            (∃p. (Ph p ∧ W m p x ∧ W m p y))
                ⟷ (space2 x y = (v * v)*(time2 x y))
        ))))"
```

```
record Body =
  Ph :: "bool"
  IOb :: "bool"

class WorldView = SpaceTime +
fixes
  (* Worldview relation *)
  W :: "Body ⇒ Body ⇒ 'a Point ⇒ bool" ("_ sees _ at _")
  ...
```

Fig. 4. A body can be a photon and/or an inertial observer. We do not require that the body should only be one or the other, because this is a theorem that can be proven from the axioms. The worldview relation is a predicate defined on two bodies and one location, and introduces the notation a sees b at x as a more intuitive rendition of W a b x. It inherits basic definitions from the class SpaceTime.

This is an essentially verbatim translation of AxPh. It assumes that various WorldView constructs of Fig. 4 are in place, including the inherited definitions of space2 and time2.

Two other first-order variants of relativity theory are also relevant here. The theory AccRel represents a kind of halfway-house: bodies can be accelerated (non-inertial), but we do not as yet include Einstein's Equivalence Principle relating acceleration to gravity. Adding an axiom representing the latter leads to GenRel, the first-order theory of general relativity. The use of the record construct in Isabelle/HOL is especially useful in this context, as it allows us to extend some of our definitions very easily. When reasoning in SpecRel, for example, we assume that bodies are either photons or inertial observers. When we come to define AccRel we can simply extend the Body record to include a third predicate for non-inertial observers, without having to re-work our earlier proof that bodies cannot be both photons and inertial observers. (Alternatively, as long as we avoid introducing a fourth type of body we can identify non-inertial observers semantically – they are bodies b for which IOb b and Ph b are both false.)

Choosing the axioms as simple as possible allows us to investigate the extent to which different axioms can be weakened without losing physical realism. For example, while AxPh says that each observer considers the speed of light to be constant, there is no assumption that different observers agree as to what this speed is (this is instead proven as a theorem). Similarly, there is no axiom declaring the sets of photons and inertial observers to be disjoint; this is another theorem. On the other hand, the drive for simplicity is not without cost. For example, the reader may be wondering why AxPh refers to the *squared* speed of light. This is because FOL is not powerful enough to characterise the field $\mathbb{R}$ of real numbers; there are fields which satisfy precisely the same first order theorems as $\mathbb{R}$ but which admit infinite values and infinitesimals [5,10]. Similarly, $\mathbb{R}$ satisfies various additional field axioms that are not always needed for the theorems we wish to prove; in particular we do not generally assume the

Euclidean axiom (that all positive quantities have square roots) because, as AxPh shows, we can redefine concepts using squared values instead. The question naturally arises, which number fields can be used when modelling relativity theory? Madarász and Székely argue that the answer depends on the underlying axiom system used to capture each particular version of relativity theory, and have demonstrated that an axiom system for special relativity can even be defined over the field $\mathbb{Q}$ of rationals [7]. Taking such considerations into account can add significantly to the work involved in stating theorems and developing their proofs.

Nonetheless, the approach has several advantages from a computational point of view. Consider, for example our Isabelle/HOL description of basic spacetime constructs. This is a 836-line file giving definitions, axioms and proofs relating to quantities, vectors, points, lines, planes and cones. This file took approximately 4 person-weeks to construct and verify, but now that it is in place the sparse nature of our assumptions and constructs means that relatively little additional work is required when moving from the special (SPECREL) to the accelerated (ACCREL) or general (GENREL) first-order theories of relativity. The main difficulty lies not in translating the underlying axioms and theorems, but in generating verifiable proofs.

2.2   Generating Verifiable Proofs

Automated theorem provers are extremely useful tools, but they are also unforgiving. For example, in our proof of Lemma 1 we wrote *the Σ_1^0 case follows by complementarity*, assuming that the reader would have sufficient mathematical competence to infer the following argument:

- if S is a Σ_1^0 set, it can be written $S = \{x \mid \exists y.R(x,y)\}$ for some recursive predicate R.
- this can be rewritten $S = \{x \mid \neg\forall y.\neg R(x,y)\}$.
- this is the complement of the set $S' = \{x \mid \forall y.\neg R(x,y)\}$.
- the predicate $R' \equiv \neg R(x,y)$ is recursive because R is recursive.
- consequently $S' = \{x \mid \forall y.R'(x,y)\}$ is a Π_1^0 set.
- consequently (as proven) S' is decidable in M-H spacetime.
- and hence $S \equiv \mathbb{N} \setminus S'$ is decidable in M-H spacetime.

Seen in this way, it is clear that the phrase *follows by complementarity* conceals a significant amount of detailed reasoning, and all of this reasoning would need to be expressed in machine-readable form if we were to attempt a machine-verification of our proof.

As our machine verification of the SPECREL theorem "no observer can travel faster than light" reveals, this problem of abbreviated reasoning is just as pronounced when discussing proofs relating to physical theories. Indeed, the bulk of the work involved choosing sensible descriptions of what we mean by geometrical terms like *line*, *plane* and *cone*. For example, while a mathematician would accept that two lines that are both parallel to a third line must be parallel to each other, this required detailed proof within Isabelle/HOL (Fig. 5).

```
lemma lemParallelTrans:
  assumes "lineA \<parallel> lineB"
    and    "lineB \<parallel> lineC"
    and    "direction lineB \<noteq> vecZero"
  shows    "lineA \<parallel> lineC"
proof -
  have case1: "direction lineA = vecZero \<longrightarrow> ?thesis" by auto
  have case2: "direction lineC = vecZero \<longrightarrow> ?thesis" by auto

  {
    assume case3: "direction lineA \<noteq> vecZero \<and> direction lineC \<noteq> vecZero"

    have exists_kab: "\<exists>kab.(kab \<noteq> (0::'a) \<and> direction lineB = kab**direction lineA)"
      by (metis parallel.simps assms(1) case3 assms(3))
    then obtain kab where kab_props: "kab \<noteq> 0 \<and> direction lineB = kab**direction lineA" by auto

    have exists_kbc: "\<exists>kbc.(kbc \<noteq> (0::'a) \<and> direction lineC = kbc**direction lineB)"
      by (metis parallel.simps assms(2) case3 assms(3))
    then obtain kbc where kbc_props: "kbc \<noteq> 0 \<and> direction lineC = kbc**direction lineB" by auto

    def kac \<equiv> "kbc * kab"
    have kac_nonzero: "kac \<noteq> 0" by (metis kab_props kac_def kbc_props no_zero_divisors)
    have "direction lineC = kac**direction lineA"
      by (metis kab_props kbc_props kac_def lemScaleScale)
    hence ?thesis by (metis kac_nonzero parallel.simps)
  }
  from this have "(direction lineA \<noteq> vecZero \<and> direction lineC \<noteq> vecZero) \<longrightarrow>
  ?thesis" by blast

  thus ?thesis by (metis case1 case2)
qed
```

Fig. 5. Isabelle/HOL proof that if two lines are both parallel to a third line, then they are also parallel to each other.

Having constructed all of the 'background' theory, translating the Hungarian proof that observers cannot travel faster than light into Isabelle/HOL form became a relatively straightforward – though still extremely time consuming – process of writing down the major steps in the proof, and then carefully filling in every possible gap in the reasoning until complete verification was achieved.

3   Next Steps

Although we have had promising results modelling SPECREL, including the first known machine verified proof of the statement "no observer can travel faster than light", the time involved in constructing these proofs means we have yet to make comparable progress developing Isabelle/HOL verification systems for theorems in ACCREL or GENREL. Our ultimate goal is to provide indisputable proof of the conjectures:

Conjecture 1. *Computation in standard Euclidean spacetime means Turing computation.*

Conjecture 2. *Computation in M-H spacetimes verifiably includes super-Turing computation.*

However, verifying these conjectures formally adds an additional layer of complexity, because they introduce a new factor not normally considered when discussing relativity theory, namely the nature of computers and computations.

In particular, as we saw in Sect. 1.1 we need to capture within Isabelle/HOL a first-order theory representing distributed computation occurring within M-H spacetimes, and we envisage having to capture a localised variant of a theory at least as complex as the π-calculus [8,11], since we need to discuss the properties of systems comprising multiple spatially-separated mobile components. Moreover, given the reliance of the schemes presented here upon the properties of M-H structures like those occurring in certain types of spacetime singularity, we will presumably also need to model what it means for a spacetime to contain a black hole, what it means for that black hole to be rotating, what it means for that rotation to be slow, and what it means for an entity to cross the event horizon. These are all new concepts in the world of Isabelle/HOL proof construction, and while we recognise that the task will require years rather than months to complete, we remain ever hopeful of eventual success.

References

1. Andréka, H., Madarász, J.X., Németi, I.: Logical analysis of relativity theories. In: Hendricks, et al. (eds.) First-Order Logic Revisited, pp. 1–30. Logos Verlag, Berlin (2004)
2. Copeland, J., Proudfoot, D.: Alan Turing's forgotten ideas in computer science. Sci. Am. **280**(4), 99–103 (1999)
3. Etesi, G., Németi, I.: Non-turing computations via Malament-Hogarth space-times. Int. J. Theor. Phys. **41**, 341–370 (2002)
4. Genzel, R., Schoedel, R., Ott, T., Eckart, A., Alexander, T., Lacombe, F., Rouan, D., Aschenbach, B.: Near-infrared flares from accreting gas around the supermassive black hole at the Galactic Centre (2003). arXiv:astro-ph/0310821
5. Goldblatt, R.: Lectures on the Hyperreals: An Introduction to Nonstandard Analysis. Graduate Texts in Mathematics, vol. 188. Springer-Verlag, Heidelberg (1998)
6. Hogarth, M.: Deciding arithmetic using SAD computers. Br. J. Philos. Sci. **55**, 681–691 (2004)
7. Madarász, J.X., Székely, G.: Special relativity over the field of rational numbers. Int. J. Theor. Phys. **52**(5), 1706–1718 (2013)
8. Milner, R.: Communicating and Mobile Systems: The Pi Calculus. Cambridge University Press, Cambridge (1999)
9. Nipkow, T.: Programming and proving in Isabelle/HOL, August 2014. http://www.cl.cam.ac.uk/research/hvg/Isabelle/dist/Isabelle/doc/prog-prove.pdf
10. Robinson, A.: Non-Standard Analysis. Princeton University Press, Princeton (1996)
11. Sangiorgi, D., Walker, D.: The Pi-Calculus: A Theory of Mobile Processes. Cambridge University Press, Cambridge (2003)
12. Siegelmann, H.: Computation beyond the Turing limit. Science **268**(5210), 545–548 (1995). http://www.dx.doi.org/10.1126/science.268.5210.545
13. Stannett, M.: Super-Turing computation. Seminar presentation, Department of computer science, University of Sheffield (1990). http://www.researchgate.net/publication/258848388_1990_Super-Turing_Computation
14. Stannett, M.: X-machines and the halting problem: building a super-Turing machine. Formal Aspects Comput. **2**(1), 331–341 (1990). http://www.dx.doi.org/10.1007/BF01888233

15. Stannett, M.: An introduction to post-Newtonian and super-Turing computation. Technical report CS-91-02, Department of Computer Science, University of Sheffield (1991). http://www.researchgate.net/publication/236852111_An_
Introduction_to_post-Newtonian_and_super-Turing_computation
16. Stannett, M.: Computation and spacetime structure. Int. J. Unconv. Comput. **9**(1–
2), 173–184 (2013)
17. Stannett, M., Németi, I.: Using Isabelle/HOL to verify first-order relativity theory.
J. Autom. Reasoning **52**(4), 361–378 (2014)
18. Welch, P.: The extent of computation in Malament-Hogarth spacetimes (2006).
arXiv:gr-qc/0609035

Regular Papers

A Connection Between Red-Green Turing Machines and Watson-Crick T0L Systems

Erzsébet Csuhaj-Varjú[1], Rudolf Freund[2], and György Vaszil[3]([✉])

[1] Department of Algorithms and Their Applications, Faculty of Informatics,
Eötvös Loránd University,
Pázmány Péter Sétány 1/c, Budapest 1117, Hungary
csuhaj@inf.elte.hu
[2] Faculty of Informatics, TU Wien, Vienna, Austria
rudi@emcc.at
[3] Department of Computer Science, Faculty of Informatics,
University of Debrecen,
P.O. Box 12, Debrecen 4010, Hungary
vaszil.gyorgy@inf.unideb.hu

Abstract. Motivated by the conceptual similarity of a *mind change* of a red-green Turing machine and of a *turn to the complementary* word in Watson-Crick L systems as well as by the fact that both red-green Turing machines and Watson-Crick L systems define infinite runs, we establish a connection between the two areas of unconventional computing. We show that the sets of numbers which can be recognized by so-called standard Watson-Crick T0L systems coincide with those recognized by red-green register machines (or red-green Turing machines). The results imply that using Watson-Crick L systems we may "go beyond Turing" in a similar way as red-green register machines and red-green Turing machines can do.

1 Introduction

Red-green Turing machines are unconventional computing devices that can "go beyond Turing" [4]. They can be considered as a type of ω-Turing machines on finite inputs with a recognition criterion based on some property of the sets of states visited infinitely (finitely) often. The set of internal states of these machines is divided into two disjoint sets, called the set of *red* states and the set of *green* states. The machine is deterministic, i.e., for any configuration there exists exactly one transition. An infinite run of the Turing machine is called *recognizing*, if and only if no red state is visited infinitely often and one or more green states are visited infinitely often. A change from a green state to a red state or reversely is called a *mind change*. It has been shown that any recursively enumerable language can be recognized by a red-green Turing machine with one mind change, and if more mind changes may take place, then the power of these constructs exceeds that of the standard Turing machines, since the complement of any recursively enumerable language can be recognized by a red-green Turing machine. The computational capacity of red-green Turing machines can be described as follows: red-green Turing machines recognize exactly the

© Springer International Publishing Switzerland 2015
J. Durand-Lose and B. Nagy (Eds.): MCU 2015, LNCS 9288, pp. 31–44, 2015.
DOI: 10.1007/978-3-319-23111-2_3

Σ_2-sets of the Arithmetical Hierarchy, and red-green Turing machines accept exactly those sets which simultaneously are Σ_2- and Π_2-sets of the Arithmetical Hierarchy. By acceptance we mean that for every word not recognized by the Turing machine the run will finally end up in red. In [1], the idea of red-green computations has been extended: firstly, variants of red-green Turing machines as red-green register (counter) machines have been introduced, secondly several variants of red-green P automata (purely communicating membrane systems working as accepting devices) have been defined. It was shown that these models simulate red-green Turing machines and thus are able to "go beyond Turing".

In this paper we discuss how the concept of a *mind change* can be related to some other concept, namely the use of Watson-Crick complementarity, from DNA computing. A notion, called a Watson-Crick D0L system (a WD0L system), where the paradigm of complementarity is considered in the operational sense, was introduced in [5]. This construct is a D0L system over a so-called DNA-like alphabet V and a mapping ϕ called the mapping defining the trigger for the complementarity transition. In a DNA-like alphabet each letter has a complementary letter and this relation is symmetric. ϕ is a mapping from the set of words (strings) over the DNA-like alphabet Σ to $\{0, 1\}$ with the following property: the ϕ-value of the axiom is 0; whenever the ϕ-value of a word is 1, then the ϕ-value of its complementary word must be 0. (The complementary word of a word is obtained by replacing each letter in the word with its complementary letter.) The derivation in a Watson-Crick D0L system is as follows: when the new word has been computed by applying the homomorphism of the D0L system, then it is checked according to the trigger. If the ϕ-value of the obtained word is 0, then the derivation continues in the usual manner; if its ϕ-value is equal to 1, then the word is changed for its complementary word and the derivation continues from this word. Thus, Watson-Crick complementarity is considered as an operation: together with or instead of a word w we consider its complementary word. Watson-Crick D0L systems have been extended to other variants of L systems, and the area has been studied in detail. Computational completeness of different variants of Watson-Crick L systems has been proved, for example in [3,9]. Decidability problems have been studied in [9], uni-transitional systems were described in [7,8], only to mention a very few articles.

Motivated by the conceptual similarity of a mind change and the concept of a turn to the complementary word as well as the fact that both red-green Turing machines and Watson-Crick L systems define infinite runs, in this paper we are going to establish a connection between these areas of unconventional computing. We first define a new notion to associate languages with Watson-Crick T0L (WT0L) systems based on the infinite word sequences that they generate: If the word sequence induced by a WT0L scheme G with the axiom w is *ultimately stable* (that is, generated using only a finite number of complementary turns), then we say that w is recognized by G. Then the connection between the complementary turns of Watson-Crick T0L systems and the mind changes of red-green Turing machines is established by showing that the sets of numbers which can be recognized by WT0L systems coincide with those recognized by red-green register machines (or red-green Turing machines). The results imply that using Watson-Crick L systems it is possible to "go beyond Turing". The obtained results shed new light on the unconventional nature of computing devices in natural computing, i.e., in DNA computing as well.

We note that the idea of "going beyond Turing" in DNA computing appeared already long time ago when in [6] the concept of computing by *carving* was introduced: to generate a set of candidate solutions of a problem, and then remove the non-solutions such that what remains is the solution. In that paper it was shown that by carving non-recursively enumerable languages can be computed.

2   Preliminaries

We assume the reader to be familiar with the basic notions of formal language and computability theory; for further details we refer to [10].

For an alphabet V, V^* denotes the set of all words over V; if λ, the empty word is not included, we use the notation V^+. For an alphabet V, $a \in V$ and $u \in V^*$, $|u|_a$ denotes the number of occurrences of a in u.

A register machine is a construct $M = (m, B, l_0, l_h, P)$, where m is the number of registers, B is the set of labels, l_0 is the initial label, l_h is the final label, and P is the set of instructions labelled by elements of B. The instructions have one of the following forms:

- $(l_i : ADD(r); l_j)$, where $l_i \in B \setminus \{l_h\}$, $l_j \in B$, $1 \le r \le m$.

  This instruction, labelled by l_i, is called an increment instruction; it increases the value of register r by one and then the computation continues with instruction l_j.
- $(l_i : SUB(r); l_j, l_k)$, where $l_i \in B \setminus \{l_h\}$, $l_j, l_k \in B$, $1 \le r \le m$.

  This instruction is called a subtract instruction. If the number stored in register r is not zero, then this instruction decreases this number by one and then the computation continues with instruction l_j; this case is called *decrement*. If the value of register r is zero, then without performing any change of the registers, the computation continues with instruction l_k; this case is called *zero-test*.
- $l_h : HALT$. This instruction stops the work of the register machine.

A configuration of the register machine is described by the current instruction label and the contents of the registers, i.e., the numbers stored in them. The current instruction label identifies the instruction to be executed. The register machine works with changing its configurations, these changes are also called transitions. A transition sequence starting with the initial instruction l_0 and ending with the final instruction l_h is called a computation by M. A natural number n is said to be accepted by M if there is a halting computation (a computation ending with instruction l_h) such that at the beginning of the computation n is stored in the first register and all other registers store the value 0.

It is well-known that for any recursively enumerable set of natural numbers there exists a register machine M that has at most three registers and accepts this set of numbers.

2.1   Red-Green Turing Machines

In [4] an extension of the notion of a Turing machine, called *red-green Turing machine* was introduced:

Definition 1. *A Turing machine TM is called a* red-green Turing machine *if its state set Q is divided into two disjoint sets: Q_{red}, the set of so-called* red *states, and Q_{green}, the set of so-called* green *states. Furthermore, TM is deterministic, i.e., for each configuration there is exactly one transition to the next one. A state p of a red-green Turing machine TM is said to be of color x, if $p \in Q_x$, where $x \in \{red, green\}$. If in a transition from configuration C to configuration C' the two corresponding states q and q' are of different color, then we speak of a* mind change. *The initial state is a red state.*

Red-green Turing machines work on finite inputs with the following acceptance criteria: An input word w is recognized by the red-green Turing machine TM if for the infinite run on w no red state is visited infinitely often and some green states are visited infinitely often. It is known (see [4]) that a set of words L is recognized by a red-green Turing machine with one mind change if and only if L is recursively enumerable. If more mind changes are allowed, the power of red-green Turing machines is revealed – red-green Turing machines recognize exactly the Σ_2-sets of the Arithmetical Hierarchy.

There are several ways to define the Arithmetical Hierarchy, we briefly refer to the following. A way to extend the hierarchy of unsolvable problems is to ask if a computer program will generate an infinite number of outputs. This property can be generalized by interpreting the output of a computer as the Gödel number of another computer. Then one can ask the question "Does a program have an infinite number of outputs an infinite subset of which, when interpreted as computer programs, have an infinite number of outputs?" This can be iterated any finite number of times to create the Arithmetical Hierarchy. In that sense, the Arithmetical Hierarchy can be described as in Table 1 taken from [2].

In the analogy to red-green Turing machines, in [1] red-green counter machines (register machines) have been introduced. In the following we recall the notion, with the necessary modifications, namely, no input word is read by the red-green register machine, and as it is common for register machines, we only deal with natural numbers (non-negative integers).

Definition 2. *A red-green register machine is a construct*

$$RM = (m, B, B_{red}, B_{green}, l_0, P)$$

where m is the number of registers, B is the set of the labels of the instructions in the instruction set P, and B is divided into two disjoint sets, B_{red} (red labels) and B_{green} (green labels). As in the case of standard register machines, l_0 is the initial label; label l_h, the halting label, can be omitted. A configuration of a red-green register machine $RM = (m, B, B_{red}, B_{green}, l_0, P)$ is denoted by $(l; r_1, \ldots, r_m)$ where $l \in B$ and $r_1, \ldots, r_m$ are the natural numbers stored in the registers.

Table 1. Arithmetical Hierarchy

Level	Question: will the computer program
$\Sigma_0 = \Pi_0$	halt in fixed time
Σ_1	ever halt
Π_1	never halt
Σ_2	have at most a finite number of outputs
Π_2	have an infinite number of outputs
Σ_3	have at most a finite number of Π_2 outputs
Π_3	have an infinite number of Π_2 outputs
Σ_n	have at most a finite number of Π_{n-1} outputs
Π_n	have an infinite number of Π_{n-1} outputs

Notice that register machines can be considered as counter machines operating on a unary alphabet; the contents of the first register at the beginning of the computation can be considered as a unary input word of the corresponding counter machine.

Similarly to red-green Turing machines, red-green register machines recognize the contents $n \in \mathbb{N}$ of their input register, if for the infinite run on n, some green state is visited infinitely often, while red states are visited only finitely often.

Throughout the paper, $N(RM)$ denotes the length sets of the unary words in the first register recognized by the red-green register machine RM; we may also say that RM computes $N(RM)$.

In [1] it was shown that the computations of a red-green Turing machine TM can be simulated by a red-green counter machine RM in such way that during the simulation of a transition of TM leading from a state p with color x to a state q with color y, for $x, y \in \{green, red\}$, the simulating counter machine uses instructions with labels of color x and only in the last step of the simulation changes to a label of color y, and reversely. That is, the simulating instruction sequence or the simulating transition sequence performs as many mind changes as the simulated transition or the simulated instruction means. It was shown that a language L is recognized by a red-green counter machine with one mind change if and only if $L \in \Sigma_1$, i.e., L is recursively enumerable, and red-green counter machines recognize exactly the Σ_2-sets of the Arithmetical Hierarchy.

2.2 Watson-Crick L Systems

By a *T0L system* we mean a construct $H = (V, g_1, \ldots, g_n, w_0)$ where V is an alphabet, the g_i, $1 \leq i \leq n$, are endomorphisms defined on V^*, and $w_0 \in V^*$ is the axiom. A word sequence σ of H is defined as a sequence of words $w_0, w_1, w_2, \ldots$ where for every $i \geq 0$ we have $w_{i+1} = g_j(w_i)$ for some $1 \leq j \leq n$; we also may say that w_i directly derives w_{i+1}. Instead of endomorphisms g_j, $1 \leq j \leq n$, we also may use the notation T_j, where T_j is a finite set of productions

of the form $a \to u$, where $a \in V$, $u \in V^*$, and for every letter $a \in V$ there is
at least one production in T_j; T_j is called a *table* of H. If the axiom, w_0, is not
indicated, then we speak of a *T0L scheme*.

In the following we recall the basic notions for *Watson-Crick L systems*.

By a DNA-like alphabet V we mean an alphabet with $2n$ letters, $n \geq 1$,
of the form $\Sigma = \{a_1, \ldots, a_n, \bar{a}_1, \ldots, \bar{a}_n\}$. Letters a_i and $\bar{a}_i$, $1 \leq i \leq n$, are
said to be complementary letters; we also call the non-barred symbols *purines*
and the barred symbols *pyrimidines*. The terminology originates from the basic
DNA alphabet $\{A, G, C, T\}$, where the letters A and G are for purines and their
complementary letters T and C are for pyrimidines.

By h_w we denote the letter-to-letter endomorphism of a DNA-like alphabet
V mapping each letter to its complementary letter; h_w is also called the *Watson-Crick morphism*.

Definition 3. *A* Watson-Crick T0L system *(a* WT0L system, *for short) is a
pair* $W = (H, \phi)$ *where* $H = (V, g_1, \ldots, g_n, w_0)$ *is a T0L system with a DNA-like
alphabet* V, *endomorphisms* g_j, $1 \leq j \leq n$, *and axiom* $w_0 \in V^+$, *and* $\phi : V^* \to \{0, 1\}$ *is a recursive function such that* $\phi(w_0) = \phi(\lambda) = 0$ *and for every word*
$u \in V^*$ *with* $\phi(u) = 1$ *it holds that* $\phi(h_w(u)) = 0$.

A word sequence σ *of a Watson-Crick T0L system* W *consists of words*
$w_0, w_1, w_2, \ldots$, *where for each* $i \geq 0$ *there exists a* k, $1 \leq k \leq n$, *such that*

$$w_{i+1} = \begin{cases} g_k(w_i) & \text{if } \phi(g_k(w_i)) = 0 \\ h_w(g_k(w_i)) & \text{if } \phi(g_k(w_i)) = 1. \end{cases}$$

The condition $\phi(u) = 1$ is said to be the trigger for the complementarity
transition. If $w_{j+1} = h_w(g_k(w_j))$, then we say that a turn to the complement
takes places when obtaining w_{j+1} from w_j. If it is clear from the context, then
we can omit the reference to ϕ.

For any pair (w_i, w_{i+1}), $i \geq 0$, we also may use the notation $w_i \Longrightarrow_G w_{i+1}$
called a computation step or a derivation step from w_i to w_{i+1} in G. The reflexive
transitive closure of relation $\Longrightarrow_G$ is denoted by $\Longrightarrow_G^*$. The word sequence σ is
an infinite computation (infinite derivation) in G.

As in the case of T0L systems, if no axiom is indicated, then we speak of a
WT0L scheme.

Various mappings ϕ are able to satisfy the conditions of defining a trigger
for the complementarity transition. In the following, we shall use a particular
variant and the corresponding WT0L system is called *standard*. In this case a
word w satisfies the trigger for turning to the complementary word if it has more
occurrences of pyrimidines (barred letters) than purines (non-barred letters).
Formally, consider a DNA-like alphabet $V = \{a_1, \ldots, a_n, \bar{a}_1, \ldots, \bar{a}_n\}$, $n \geq 1$. Let
$V_{PUR} = \{a_1, \ldots, a_n\}$ and $V_{PYR} = \{\bar{a}_1, \ldots, \bar{a}_n\}$. Then, we define $\phi : V^* \to \{0, 1\}$
as follows: for $w \in V^*$

$$\phi(w) = \begin{cases} 0 \text{ if } |w|_{V_{PUR}} \geq |w|_{V_{PYR}} \text{ and} \\ 1 \text{ if } |w|_{V_{PUR}} < |w|_{V_{PYR}}. \end{cases}$$

We note that in the case of standard Watson-Crick L systems the trigger ϕ
is given by a context-free context condition.

Definition 4. *A word sequence σ of a WT0L system W is said to be* ultimately stable *if there exists a word w_i in the sequence such that for any $j \geq i$, $w_{j+1} = g_k(w_j)$ for some k, $1 \leq k \leq n$.*

Notice that if σ is ultimately stable, then there are only a finite number of turns to the complement in generating the elements of σ.

Languages can be associated with WT0L schemes in different ways; in the following, we introduce a new concept, in accordance with the notions related to red-green Turing machines.

Definition 5. *We say that a WT0L scheme $G = (V, g_1, \ldots, g_n, \phi)$ recognizes a word $w \in V^*$ if there is a word sequence of the WT0L system $W = (V, g_1, \ldots, g_n, w, \phi)$ which is ultimately stable.*

3   Results

We show that transition sequences of red-green register machines (Turing machines) can be simulated by computations with standard WT0L schemes and vice versa.

Lemma 1. *Let $RM = (m, B, B_{red}, B_{green}, P)$, $m \geq 1$, be a red-green register machine. Then there exists a standard WT0L scheme $G = (V, T_1, \ldots, T_n, \phi)$, $n \geq 1$, such that for every transition in RM of the form*

$$(l_i, r_1, \ldots, r_m) \Longrightarrow_{RM} (l_j, r_1', \ldots, r_m')$$

and for every $d \in V^*$ where $|d|_{a_i} - |d|_{\bar{a}_i} = r_i$, it holds that there exists a derivation

$$l_i d \Longrightarrow_G^* l_j d'$$

such that $|d'|_{a_i} - |d'|_{\bar{a}_i} = r_i'$, and if $l_i, l_j \in B_x$, $x \in \{red, green\}$, then during the derivation $l_i d \Longrightarrow_G^* l_j d'$ in G there is no turn to the complement, and, on the other hand, if $l_i \in B_x$ and $l_j \in B_y$ for $x \neq y$, then there is exactly one turn to the complement during the derivation $l_i d \Longrightarrow_G^* l_j d'$ in G.

Proof. Suppose that RM has s instructions, i.e., $card(P) = s$ holds. Recall that RM is deterministic, i.e., for any label $l \in B$ there is exactly one instruction labelled by l. Since each transition is realized by executing a certain instruction, to prove the statement we construct tables of the WT0L scheme which being applied to words representing the corresponding configuration simulate the effect of the execution of an instruction.

We construct the simulating WT0L scheme G as follows: Let

$$V = V_r \cup V_r' \cup Lab \cup \{\#, \bar{\#}\},$$
$$V' = V_r \cup V_r' \cup Lab,$$
$$V_r = \{a_i, \bar{a}_i \mid 1 \leq i \leq m\},$$
$$V_r' = \{b_i, \bar{b}_i \mid 0 \leq i \leq m\},$$
$$Lab = \{l, \bar{l}, l', \bar{l}', l'', \bar{l}'' \mid l \in B\}.$$

Now we define the tables of G; we present these tables together with explanations of their role in the derivations; further technical details of the proof are left to the reader. In the tables, if for some symbol no rule is indicated, then the identical rule is implicitly assumed to be taken.

We first provide an auxiliary table which helps us to induce an infinite number of turns to the complement in G; this table

$$T_\# = \{\# \to \bar{\#}\bar{\#}\#\#, \; \bar{\#} \to \#\#\#\#\}$$

will be part of every table of G. The symbols $\#$ and $\bar{\#}$ are trap symbols, and once a symbol $\bar{\#}$ has been introduced, the rule $\bar{\#} \to \#\#\#\#$ guarantees that the number of trap symbols $\#$ at some moment must exceed the number of non-barred other symbols (as the number of symbols on the right-hand sides of any other production does not exceed three), which triggers the complementarity transition in the next step. With its "twin rule" $\# \to \bar{\#}\bar{\#}\#\bar{\#}$ in the next step the same situation must arrive again which will become clear from the description of the other tables in G which perform the simulation of the instructions of RM.

Any transition of RM corresponds to the execution of one of its instructions; therefore, for every instruction we define a set of tables of G which simulates its execution. We discuss the cases of increment and subtract instructions separately, depending also on whether or not they represent a mind change or not. The idea of getting the table for the mind changing case from the other case mainly is to convert all non-barred symbols to barred symbols, which means that if before the non-barred symbols had the majority, afterwards the barred symbols will have the majority and thus cause a complementarity transition in the next step.

Another important feature of the construction is that the number stored in the register r is not represented as the corresponding number of symbols a_r, but as the difference between the number of symbols a_r and $\bar{a}_r$. This trick is needed to perform the decrement case of a subtract instruction on register r by adding a barred version $\bar{a}_r$ of a_r instead of erasing one copy of a_r, which has the same effect when taking the right interpretation as defined above.

Let $(l_i : ADD(r); l_j)$, $1 \leq i, j \leq s$, $1 \leq r \leq m$, be an increment instruction in RM such that $l_i, l_j \in B_x$, where $x \in \{red, green\}$. The corresponding table of G is defined as follows:

$$\begin{aligned} T_{l_i} = & \{l_i \to l_j a_r\} \cup \{l \to \# \mid l \in Lab, l \neq l_i\} \\ & \cup \{a_t \to a_t, \bar{a}_t \to \bar{a}_t \mid 1 \leq t \leq m\} \\ & \cup \{c \to c \mid c \in V_r'\} \cup T_\#. \end{aligned}$$

Notice the role of letters $c \in V_r'$ – those symbols which are not affected by the instruction rules given in the first two lines of the description of T_{l_i} remain unchanged. Applying this table T_{l_i}, l_i is changed for l_j and one more occurrence of a_r is added to the word in this derivation step, i.e., the instruction is simulated. If the table is applied to a word containing an occurrence of $l \in Lab, l \neq l_i$, then $\#$ is introduced which means that the obtained word does not represent any configuration of RM, and from this point on the rules from $T_\#$ will lead to an infinite sequence of turns to the complement as already argued above. Notice that the application of T_{l_i} does not imply any turn to the complement.

We now discuss the case of an increment instruction $(l_i : ADD(r); l_j)$ with a mind change, i.e., $l_i \in B_x$ and $l_j \in B_y$ with $x, y \in \{red, green\}$ and $x \neq y$. The corresponding table in G simulating this instruction is the following (superscript mc refers to mind change):

$$T_{l_i}^{mc} = \{l_i \rightarrow \bar{l}_j \bar{a}_r\} \cup \{l \rightarrow \bar{\#} \mid l \in Lab, l \neq l_i\}$$
$$\cup \{a_t \rightarrow \bar{a}_t, \bar{a}_t \rightarrow a_t \mid 1 \leq t \leq m\}$$
$$\cup \{c \rightarrow c \mid c \in V_r'\} \cup T_{\#}.$$

It is easy to see that after applying the rules of $T_{l_i}^{mc}$ a turn to the complement should be performed. Otherwise, the table simulates the execution of the instruction, in the same manner as described in the previous case.

Now we turn to the case of subtract instructions, i.e., let $(l_i : SUB(r); l_j, l_k)$, $1 \leq i, j, k \leq s$, be a subtract instruction in RM such that $l_i \in B_x$ and $l_j \in B_y$ with $x, y \in \{red, green\}$.

For this subtract instruction, we first consider the decrement case, namely, the case, when RM decreases the value of register r by 1 and then continues the computation with the instruction labelled by l_j.

If $x = y$, then the corresponding tables of G are given as follows:

$$T_{l_i,1} = \{l_i \rightarrow l_i'' \bar{b}_0 \bar{a}_r\} \cup \{l \rightarrow \bar{\#} \mid l \in Lab, l \neq l_i\}$$
$$\cup \{a_r \rightarrow a_r, \bar{a}_r \rightarrow \bar{a}_r\} \cup \{a_t \rightarrow a_t \bar{b}_t, \bar{a}_t \rightarrow \bar{a}_t b_t \mid 1 \leq t \leq m, t \neq r\}$$
$$\cup \{c \rightarrow c \mid c \in V_r'\} \cup T_{\#},$$
$$T_{l_i,2} = \{l_i'' \rightarrow l_j\} \cup \{l \rightarrow \bar{\#} \mid l \in Lab, l \neq l_i''\}$$
$$\cup \{a_t \rightarrow a_t, \bar{a}_t \rightarrow \bar{a}_t \mid 1 \leq t \leq m\}$$
$$\cup \{c \rightarrow \lambda \mid c \in V_r'\} \cup T_{\#}.$$

If register r is not empty, then the number of symbols a_r in the word is greater than the number of symbols $\bar{a}_r$. Then, after applying $T_{l_i,1}$, the word does not turn to the complement, and by applying $T_{l_i,2}$, the "assistant" letters b and $\bar{b}$ disappear. If this is not the case, i.e., if register r is empty, which means that the number of symbols a_r in the word equals the number of symbols $\bar{a}_r$, then the complementarity transition will take place, the symbol $\bar{l}_i'$ will show up in the word and then table $T_{l_i,2}$ (and any other table constructed in this proof for G) will introduce an occurrence of $\#$, which finally will lead to an infinite run with infinitely many mind changes due to the rules in $T_{\#}$ as already explained above.

If $x \neq y$, then the corresponding tables of G are obtained by modifying $T_{l_i,1}$ and $T_{l_i,2}$:

$$T_{l_i,1}^{mc} = \{l_i \rightarrow \bar{l}_i' a_r\} \cup \{l \rightarrow \bar{\#} \mid l \in Lab, l \neq l_i\}$$
$$\cup \{a_r \rightarrow \bar{a}_r, \bar{a}_r \rightarrow a_r\} \cup \{a_t \rightarrow \bar{a}_t b_t, \bar{a}_t \rightarrow a_t \bar{b}_t \mid 1 \leq t \leq m, t \neq r\}$$
$$\cup \{c \rightarrow c \mid c \in V_r'\} \cup T_{\#},$$
$$T_{l_i,2}^{mc} = \{l_i' \rightarrow l_j\} \cup \{l \rightarrow \bar{\#} \mid l \in Lab, l \neq l_i'\}$$
$$\cup \{a_t \rightarrow a_t, \bar{a}_t \rightarrow \bar{a}_t \mid 1 \leq t \leq m\}$$
$$\cup \{c \rightarrow \lambda \mid c \in V_r'\} \cup T_{\#}.$$

If the value of register r is not zero, i.e., if there are more symbols a_r in the word than symbols $\bar{a}_r$, then after applying $T_{l_i,1}^{mc}$, the word has to perform a change to the complement. Then we use table $T_{l_i,2}^{mc}$ to get l_j and for deleting the "assistant" letters from V_r' .

If the guess that register r stores a number greater than zero was wrong, then no turn to the complement follows after applying $T_{l_i,1}^{mc}$. In this case, the symbol $\bar{l}_i'$ remains in the word, but in all tables constructed in this proof, we only find the rule $\bar{l}_i' \to \bar{\#}$ thus introducing the trap symbol $\bar{\#}$ in the word.

We finish the discussion with the case when we assume that the value stored in register r is zero and RM continues its work with instruction label l_k. Register r storing zero means that in the corresponding word in G the number of symbols a_r equals the number of symbols $\bar{a}_r$. Again we have to distinguish between the two subcases $x = y$ and $x \neq y$, where $l_i \in B_x$, $l_k \in B_y$, $x, y \in \{red, green\}$.

Let first $x = y$. We define

$$
\begin{aligned}
T_{l_i,3} = &\{l_i \to l_i''\bar{b}_0\} \cup \{l \to \bar{\#} \mid l \in Lab, l \neq l_i\} \\
&\cup \{a_r \to \bar{a}_r, \bar{a}_r \to a_r\} \cup \{a_t \to a_t \bar{b}_t, \bar{a}_t \to \bar{a}_t b_t \mid 1 \leq t \leq m, t \neq r\} \\
&\cup \{c \to c \mid c \in V_r'\} \cup T_\#, \\
T_{l_i,4} = &\{l_i'' \to l_k\} \cup \{l \to \bar{\#} \mid l \in Lab, l \neq l_i''\} \\
&\cup \{a_r \to \bar{a}_r, \bar{a}_r \to a_r\} \cup \{a_t \to a_t, \bar{a}_t \to \bar{a}_t \mid 1 \leq t \leq m, t \neq r\} \\
&\cup \{c \to \lambda \mid c \in V_r'\} \cup T_\#.
\end{aligned}
$$

After applying table $T_{l_i,3}$, no turn to the complement takes place if and only if the number of symbols a_r in the word equals the number of symbols $\bar{a}_r$ (note that these numbers also can be zero, but the difference between the number of symbols a_r and the number of symbols $\bar{a}_r$ is always non-negative), i.e., if and only if register r stores zero. Otherwise, after a complementarity transition, the symbol $\bar{l}_i''$ would appear, which can only go to the trap symbol $\bar{\#}$ in any table.

Then we use table $T_{l_i,4}$ to get l_k, to recover a_r from $\bar{a}_r$ and vice versa as well as for deleting the "assistant" letters b and $\bar{b}$.

If the number of symbols a_r in the word does not equal the number of symbols $\bar{a}_r$, then a complementarity transition takes place and the symbol $\bar{l}_i''$ we obtain can only go to the trap symbol $\bar{\#}$.

Let us finally consider the case $x \neq y$. We then define

$$
\begin{aligned}
T_{l_i,3}^{mc} = &\{l_i \to \bar{l}_i''\} \cup \{l \to \bar{\#} \mid l \in Lab, l \neq l_i\} \\
&\cup \{a_r \to a_r, \bar{a}_r \to \bar{a}_r\} \cup \{a_t \to \bar{a}_t b_t, \bar{a}_t \to a_t \bar{b}_t \mid 1 \leq t \leq m, t \neq r\} \\
&\cup \{c \to c \mid c \in V_r'\} \cup T_\#, \\
T_{l_i,4}^{mc} = &\{l_i'' \to l_k\} \cup \{l \to \bar{\#} \mid l \in Lab, l \neq l_i''\} \\
&\cup \{a_r \to \bar{a}_r, \bar{a}_r \to a_r\} \cup \{a_t \to a_t, \bar{a}_t \to \bar{a}_t \mid 1 \leq t \leq m, t \neq r\} \\
&\cup \{c \to \lambda \mid c \in V_r'\} \cup T_\#.
\end{aligned}
$$

The application of table $T_{l_i,3}^{mc}$ now guarantees a turn to the complement, i.e., the obtained new word should change to its complementary word, if and only if the number of symbols a_r in the word equals the number of symbols $\bar{a}_r$.

Then we apply table $T_{l_i,4}^{mc}$ to get l_k and to recover a_r from $\bar{a}_r$ and vice versa as well as for deleting the "assistant" letters b and $\bar{b}$.

The standard WT0L scheme G contains exactly all those tables described above. It is easy to see that the tables can be applied only in the correct order, and by the explanations given with the definitions of the tables, it can be seen that the simulation result holds.   □

Based on the preceding lemma we can establish a connection between sets of numbers recognized by red-green register machines and sets of numbers recognized by standard WT0L schemes. We show that for any red-green register machine RM there exists a standard WT0L scheme G such that the set of natural numbers recognized by RM can be computed by G.

Theorem 1. *Let k be a natural number and let $RM = (m, B, B_{red}, B_{green}, P)$ be a red-green register machine which recognizes k. Then there exists a standard WT0L scheme $G = (V, T_1, \ldots, T_n, \phi)$, where $B \subseteq V$, $\{a_1 \ldots, a_m\} \subseteq V$, such that G recognizes $l_0 a_1^k a_2^0 \ldots a_m^0$.*

Proof. By Lemma 1 we can construct a standard WT0L system G such that G simulates the computations in RM, namely for any transition of RM of the form $(l_i, r_1, \ldots, r_m) \Longrightarrow_{RM} (l_j, r_1', \ldots, r_m')$ and for any $d \in V^*$ where $|d|_{a_i} - |d|_{\bar{a}_i} = r_i$, $|d'|_{a_i} - |d'|_{\bar{a}_i} = r_i'$, there exists a derivation

$$l_i d \Longrightarrow_G^* l_j d'$$

and, moreover, if $l_i, l_j \in B_x$, $x \in \{red, green\}$, then during the derivation $l_i d \Longrightarrow_G^* l_j d'$ in G there is no turn to the complement, and, on the other hand, if $l_i \in B_x$ and $l_j \in B_y$ for $x \neq y$, then there is exactly one turn to the complement during the derivation $l_i d \Longrightarrow_G^* l_j d'$ in G.

The computation in RM starts with a configuration $(l_0, r_1, 0, \ldots, 0)$ and after a while it enters a configuration $(l_p, r_1', \ldots, r_m')$ such that $l_p \in B_{green}$, and from this time on no mind change takes place any more. We construct all tables as given in the proof of Lemma 1. Then, starting a derivation from $l_0 a_1^{r_1} a_2^0 \ldots a_m^0$ and simulating the corresponding recognizing computation in RM, we will obtain the word $l_p u$ where u is a permutation of the word $a_1^{r_1''} \ldots a_m^{r_m''} \bar{a}_1^{\bar{r}_1''} \ldots \bar{a}_m^{\bar{r}_m''}$ with $r_i' = r_i'' - \bar{r}_i''$, $1 \leq i \leq m$, thus establishing an ultimately stable computation in G. This means that G recognizes $l_0 a_1^{r_1} a_2^0 \ldots a_m^0$, i.e., it computes the natural number r_1. This implies that the statement holds.   □

Definition 6. *Let $G = (V, T_1, \ldots, T_n, \phi)$ be a WT0L scheme and let $a \in V$. Then we define*

$$N_a(G) = \{k \mid w \text{ is recognized by } G, |w|_a = k \text{ and } |w|_{\bar{a}} = 0\}.$$

The following statement is a direct consequence of Theorem 1.

Corollary 1. *For any red-green register machine RM there exist a standard WT0L scheme $G = (V, T_1, \ldots, T_n, \phi)$ and an $a \in V$ such that $N(RM) = N_a(G)$.*

Next we will demonstrate how sequences of transitions of red-green Turing machines can simulate derivations of standard WT0L schemes.

Lemma 2. *Let $G = (V, T_1, \ldots, T_r, \phi)$, $r \geq 1$, be a standard WT0L scheme. Then there exist a red-green Turing machine TM and two states $p \in Q_{red}$ and $q \in Q_{green}$ such that for any $u, v \in V^*$ with $u \Longrightarrow_G v$ there exist two computations c_1, c_2 in TM, such that*

- *computation c_1 starts in p and computation c_2 starts in q,*
- *both computations obtain v from u,*
- *if $u \Longrightarrow_G v$ involved a turn to the complement, then c_1 ends in q and c_2 ends in p, otherwise, c_1 ends in p and c_2 ends in q.*

Proof. We construct the red-green Turing machine TM as follows: the input word u is put on the input tape in the form $\$_1 u \$_2$, i.e., between two markers, and the machine is in either of the states p or q. We describe the computation from state p, the other state can be treated in a similar manner. Until the last step, the states will have the same "color" as p, i.e., red. TM reads the input from left to right, starting with the first letter following $\$_1$, and for each letter a in u it writes a word u_a on a second worktape, where $a \to u_a$ is a rule in some table of G. During the whole procedure, i.e., until reaching $\$_2$, TM uses rules of the same table of G. Now the word obtained on the worktape is copied to the input tape between the two markers, and it is read again from left to right: using two other worktapes, TM checks whether or not the barred letters are in majority in the obtained word. The worktapes work as counters: if a non-barred letter is read, then the machine adds one symbol to the contents of the first tape, if a barred letter is scanned, then it adds a symbol to the contents of the second worktape. When TM reaches $\$_2$, then it makes a comparison of the contents of the two worktapes and enters state s if the number of non-barred letters is equal or greater than the barred letters, or else enters state $\bar{s}$, in both cases still in a state with color red. Now TM "cleans" all worktapes and returns its reading head to the position of $\$_1$, using different red states to end up with s' when having started this final procedure from state s and with $\bar{s}'$ otherwise. Finally, TM enters p from s', i.e., if no mind change takes place, or else q from $\bar{s}'$ otherwise, i.e., if a mind change has to take place.

It can easily be seen that TM performs one mind change if and only if G performs a turn to the complement.   □

As a consequence of the preceding statement we obtain the following theorem.

Theorem 2. *Any language that can be recognized by a standard WT0L scheme can be recognized by a red-green Turing machine.*

Proof. By the preceding lemma, any derivation step of a standard WT0L scheme G can be simulated by some transitions of a red-green Turing machine TM. Thus, a derivation in G can also be simulated by a sequence of transition sequences in TM. Furthermore, the simulation preserves mind change, i.e., whenever a turn to the complement takes place in G, then a mind change takes place in the simulating transition sequence in TM. Thus, it can be seen that recognizing a derivation by G corresponds to recognizing a computation by TM.   □

Corollary 2. *Any set of natural numbers that can be computed by a standard WT0L scheme can be computed by a red-green register machine and vice versa.*

4   Conclusions

We have demonstrated the relationship of ideas from different fields of unconventional computing: Watson-Crick T0L systems and red-green register machines or red-green Turing machines. We first introduced a way to associate languages with Watson-Crick T0L (WT0L) systems based on the infinite word sequences that they generate: If the word sequence induced by a WT0L scheme G with the axiom w is ultimately stable (that is, generated using only a finite number of complementary turns), then we say that w is recognized by G. The connection between the complementary turns of Watson-Crick T0L systems and the mind changes of red-green Turing machines is established by showing that the sets of numbers which can be recognized by WT0L systems coincide with those recognized by red-green register machines or red-green Turing machines.

To demonstrate the power of WT0L systems, we repeat an argument from [4] to show how the complement of any recursively enumerable language can be characterized by a WT0L system. As the complements of recursively enumerable languages are not necessarily recursively enumerable, this shows that (similarly to red-green Turing machines) the language recognizing power of WT0L systems is greater than the power of standard Turing machines. To recognize the complement of a given recursively enumerable language L, we first consider the register machine machine M with $L(M) = L$, that is, the machine that halts by executing a specific halt instruction when started with $w \in L$ in its first register, and executes an infinite computation when started with $w \notin L$ in the first register. Based on the technique presented in the proof of Lemma 1, we can construct a WT0L system G which simulates M. Now, if we also add rules to G in such a way that after simulating the halting instruction of M, it enters an infinite cycle turning to the complement in each step, we obtain a WT0L system G' where the only ultimately stable word sequences are generated by computations which simulate the infinite (and therefore non-accepting) computations of the register machine M, which means that the language recognized by G' is exactly the complement of L, the language accepted by M.

References

1. Aman, B., Csuhaj-Varjú, E., Freund, R.: Red–green P automata. In: Gheorghe, M., Rozenberg, G., Salomaa, A., Sosík, P., Zandron, C. (eds.) CMC 2014. LNCS, vol. 8961, pp. 139–157. Springer, Heidelberg (2014)
2. Budnik, P.: What Is and What Will Be. Mountain Math Software, Los Gatos (2006)
3. Csima, J., Csuhaj-Varjú, E., Salomaa, A.: Power and size of extended Watson-Crick L systems. Theor. Comput. Sci. **290**, 1665–1678 (2003)
4. van Leeuwen, J., Wiedermann, J.: Computation as an unbounded process. Theor. Comput. Sci. **429**, 202–212 (2012)
5. Mihalache, V., Salomaa, A.: Language-theoretic aspects of DNA complementarity. Theor. Comput. Sci. **250**, 163–178 (2001)
6. Păun, Gh.: (DNA) Computing by carving. Soft Comput. **3**, 30–36 (1999)

7. Salomaa, A.: Uni-transitional Watson-Crick D0L systems. Theor. Comput. Sci. **281**(1–2), 537–553 (2002)
8. Salomaa, A., Sosík, P.: Watson-Crick D0L systems: the power of one transition. Theor. Comput. Sci. **301**, 187–200 (2003)
9. Sosík, P.: Watson-Crick D0L systems: generative power and undecidable problems. Theor. Comput. Sci. **306**(1–2), 101–112 (2003)
10. Rozenberg, G., Salomaa, A. (eds.): Handbook of Formal Languages, vol. 1–3. Springer, Berlin (1997)

Tight Bounds for Cut-Operations on Deterministic Finite Automata

Frank Drewes[1], Markus Holzer[2]([✉]), Sebastian Jakobi[2],
and Brink van der Merwe[3]

[1] Department of Computing Science, Umeå University, Umeå, Sweden
drewes@cs.umu.se
[2] Institut für Informatik, Universität Giessen, Arndtstr. 2, 35392 Giessen, Germany
{holzer,sebastian.jakobi}@informatik.uni-giessen.de
[3] Department of Mathematical Sciences, Computer Science Division,
University of Stellenbosch, Stellenbosch, South Africa
abvdm@cs.sun.ac.za

Abstract. We investigate the state complexity of the cut and iterated cut operation for deterministic finite automata (DFAs), answering an open question stated in [M. BERGLUND, et al.: Cuts in regular expressions. In *Proc. DLT*, LNCS 7907, 2011]. These operations can be seen as an alternative to ordinary concatenation and Kleene star modelling leftmost *maximal* string matching. We show that the cut operation has a matching upper and lower bound of $(n-1) \cdot m + n$ states on DFAs accepting the cut of two individual languages that are accepted by n- and m-state DFAs, respectively. In the unary case we obtain $\max(2n-1, m+n-2)$ states as a tight bound. For accepting the iterated cut of a language accepted by an n-state DFA we find a matching bound of $1 + (n+1) \cdot \mathsf{F}(1, n+2, -n+2; n+1 \mid -1)$ states on DFAs, where F refers to the generalized hypergeometric function. This bound is in the order of magnitude $\Theta((n-1)!)$. Finally, the bound drops to $2n-1$ for unary DFAs accepting the iterated cut of an n-state DFA and thus is similar to the bound for the cut operation on unary DFAs.

1 Introduction

The equivalence of finite automata and regular expressions is well known, and appropriate constructions for the conversion between these representations of regular languages can be found in almost all monographs on automata and formal languages. Although the concepts are the same, the implementation of regular expression matching engines may result in fundamentally different finite state devices. Besides using deterministic or nondeterministic finite automata as string matchers, the main difference is their performance characteristics and operational semantics when performing the string matching of the input word against the constructed matcher. Recently, the behaviour of nondeterministic matchers was investigated in detail with respect to exponential matching time, also referred to as catastrophic backtracking [2,6]. One possibility to control the

© Springer International Publishing Switzerland 2015
J. Durand-Lose and B. Nagy (Eds.): MCU 2015, LNCS 9288, pp. 45–60, 2015.
DOI: 10.1007/978-3-319-23111-2_4

work-flow of the matcher is to use operations that prevent backtracking, similarly as in the logic programming language PROLOG [3]. In fact, language operations with such a behaviour were recently introduced in [1] as an alternative to ordinary concatenation and Kleene star modelling leftmost *maximal* string matching. In order to explain the behaviour of these new regularity preserving operations consider the following pseudo-code example, which is literally taken from [1]—and assume that match_regex matches the *longest* prefix possible:

```
match = match_regex("(a*b)*", s);
if(match != null) then
    match = match_regex("ab*c", match.string_remainder);
    if(match != null) then
        return match.string_remainder == "";
return false;
```

For the string $s = abac$, this program first matches $R = (a^*b)^*$ to the sub-string ab, leaving ac as a remainder, which is matched by $S = ab^*c$, returning the empty string as a remainder, indicating a positive match. On the other hand, for $s = aababc$ in an execution of the program above, regular expression R matches $aabab$, leaving the remainder c, which *cannot* be matched by S, thus returning false, although s belongs to $R \cdot S$. Exactly this behaviour on leftmost maximal matching is modelled by the cut and iterated cut operation. In [1] basic properties of these operations with respect to formal languages and computational complexity were investigated. In particular, both operations preserve regularity. One of the many open questions stated in [1] is to develop a better or complete understanding of the upper and lower bounds on the state complexity of finite automata for both variants of cut operations. We solve this question by giving exact matching upper and lower bounds in the number of states for deterministic finite automata (DFAs) accepting the cut of two languages or the iterated cut of a single language.

In the next section we introduce the necessary notation on DFAs. Moreover, the cut and iterated cut operation is defined and the basic automata constructions for both cut operations on languages are recalled. Then in Sect. 3 the state complexity of the cut-operation on DFAs in general and on unary DFAs is investigated. Both bounds are polynomial in n and m. To be more precise, $(n - 1) \cdot m + n$ states are sufficient and necessary to accept the cut of two languages accepted by n- and m-state DFAs, and $\max(2n - 1, m + n - 2)$ states are sufficient and necessary for *unary* DFAs. Here a DFA is unary if it has a singleton input alphabet. The tight bound for general regular languages is best possible, since the lower bound even holds for languages over a two letter alphabet. The iterated cut operation is studied in Sect. 4. Here the situation is much more involved. For DFAs in general we obtain a sufficient and necessary bound of $1 + (n+1) \cdot \mathsf{F}(1, n+2, -n+2; n+1 \mid -1)$ on the exact number of states, where F refers to the generalized hypergeometric function. It is shown that this bound is in the order of $\Theta((n-1)!)$. In the unary case the bound drops to $2n - 1$. Observe that the lower bound for the iterated cut operation for regular languages in general even holds for languages over a three letter alphabet. Whether a bound in the order of $\Theta((n-1)!)$ can already be obtained by a language over a two letter alphabet is left open. Moreover, for all presented results we also discuss the effect of the number of accepting states in the involved automata to the upper and lower bounds for the cut operations. Finally, we summarize our results in the concluding section and state some open problems for future research. Owing to space constraints some proofs had to be shortened or left out. Complete proofs will be given in a forthcoming journal version of the paper.

2 Preliminaries

We recall some definitions on finite automata as contained in [5]. A *deterministic finite automaton* (DFA) is a quintuple $A = (Q, \Sigma, \delta, q_0, F)$, where Q is the finite set of *states*, Σ is the finite set of *input symbols*, $q_0 \in Q$ is the *initial state*, $F \subseteq Q$ is the set of *accepting states*, and $\delta \colon Q \times \Sigma \to Q$ is the *transition function*. The *language accepted* by the DFA A is defined as $L(A) = \{\, w \in \Sigma^* \mid \delta(q_0, w) \in F \,\}$, where the transition function is recursively extended to $\delta \colon Q \times \Sigma^* \to Q$. A DFA is *unary*, if the input alphabet Σ is a singleton set, that is, $\Sigma = \{a\}$, for some input symbol a.

In [1] the *cut operation* on two languages L an L', denoted by $L \,!\, L'$, is defined as

$$L \,!\, L' = \{\, uv \mid u \in L, \, v \in L', \text{ and } uv' \notin L, \text{ for every } v' \in \mathrm{pref}(v) \,\},$$

where $\mathrm{pref}(v)$ denotes the set of all nonempty prefixes of the word v. Moreover, also an iterated version of the cut operation was defined. The *iterated cut* of a language L, denoted by $L^{!*}$, is the smallest language that satisfies

$$\{\lambda\} \cup (L \,!\, (L^{!*})) \subseteq L^{!*},$$

i.e., $L \,!\, (L \,!\, \ldots (L \,!\, (L \,!\, \{\lambda\})) \ldots) \subseteq L^{!*}$ for any number of repetitions of the cut. In other words, the language $L^{!*}$ is the least fixed point of $X \mapsto \{\lambda\} \cup (L \,!\, X)$. We also define $L^{!+}$ as the smallest language that satisfies $L \cup (L \,!\, (L^{!+})) \subseteq L^{!+}$, or equivalently, as least fixed point of $X \mapsto L \cup (L \,!\, X)$. Notice that $L^{!*} = L^{!+} \cup \{\lambda\}$. The above defined cut operations preserve regularity as shown in [1]. Since we are interested in the descriptional complexity of both operations we briefly recall both constructions from [1]—we slightly adapt these constructions such that they also work in case the initial state of the automaton has incoming transitions and is a possible final state. We start recalling the construction for the cut operation.

Let $A = (Q_A, \Sigma, \delta_A, q_{0,A}, F_A)$ and $B = (Q_B, \Sigma, \delta_B, q_{0,B}, F_B)$ be two DFAs accepting the languages L and L', respectively. Then define the automaton $C = (Q, \Sigma, \delta, q_0, F)$, with state set $Q = Q_A \cup Q_A Q_B$. The idea behind δ is to let C first run A and then, as soon as A has accepted a prefix of the input, both A and B in parallel, so that B can be reset to its initial state each time A encounters another (longer) prefix in L. Therefore, for all states $q_A, r_A \in Q_A$, $q' = q_A q_B \in Q$, and inputs $a \in \Sigma$ with $\delta_A(q_A, a) = r_A$ we define

$$\delta(q_A, a) = \begin{cases} r_A & \text{if } r_A \notin F_A \\ r_A \, q_{0,B} & \text{otherwise} \end{cases}$$

and

$$\delta(q', a) = \begin{cases} r_A\, \delta_B(q_B, a) & \text{if } r_A \notin F_A \\ r_A\, q_{0,B} & \text{otherwise} \end{cases}$$

and $q_0 = q_{0,A}$, if $\lambda \notin L(A)$, and $q_0 = q_{0,A}\, q_{0,B}$, otherwise. The set of final states is set to $F = Q_A F_B$. Then $L(C) = L\,!\,L'$. Since the states of C are non-empty sequences of length at most two, we refer to an element $q \in Q$ as a *stack of states* or a *stack state*. The *height* of a stack state is the length of its sequence of states. This view on the state set is used in the iterated cut construction, which is more subtle.

Again, let $A = (Q_A, \Sigma, \delta_A, q_{0,A}, F_A)$ be a DFA accepting the language L. Before we define the DFA $C = (Q, \Sigma, \delta, q_0, F)$ that accepts $L^{!+}$, we need some prerequisites in order to keep the presentation of C simple. The idea for the construction of C is as follows: first the automaton behaves like A. If it reaches one of the final states of A, say q_1, it continues in a state $q_1 q_{0,A}$, working essentially like the automaton for the language $L(A)\,!\,L(A)$. In particular, it resets the second copy each time the first copy encounters a final state of A. However, if the second copy reaches a final state q_2 of A, while $q_1 \notin F$, a third copy is initialized, thus resulting in a state of the form $q_1 q_2 q_{0,A}$, and so on. In order to keep the set of states finite we need a function $\pi : Q_A^+ \to Q_A^{\leq |Q_A|}$, which is defined as follows: for all $s = q_1 q_2 \ldots q_k \in Q_A^+$: if $k = 1$, then $\pi(s) = s$, and if $k > 1$, then

$$\pi(s) = \begin{cases} \pi(q_1 q_2 \ldots q_{k-1}) & \text{if } q_k \in \{q_1, q_2, \ldots, q_{k-1}\} \\ \pi(q_1 q_2 \ldots q_{k-1}) q_k & \text{otherwise.} \end{cases}$$

Obviously, function π removes repeated states in the state sequence from right to left. Hence, the set $\pi(Q_A^+)$ consists only of those sequences, where every state appears at most once. Now we are ready to describe C. Let $Q = \pi(Q_A^+)$ and $q_0 = q_{0,A}$. As in the previous cut construction, the elements of Q are called *stacks of states* from Q_A. Then for every $q \in Q$ with $q = q_1 q_2 \ldots q_k$ and $a \in \Sigma$ let $q_i' = \delta_A(q_i, a)$, for $1 \leq i \leq k$, and set

$$\delta(q, a) = \begin{cases} \pi(q_1' q_2' \ldots q_k') & \text{if } q_1', q_2', \ldots, q_k' \notin F_A \\ \pi(q_1' q_2' \ldots q_\ell' q_{0,A}) & \text{if } \ell = \min\{i \mid 1 \leq i \leq k \text{ and } q_i' \in F_A.\} \end{cases}$$

The set of final states is

$$F = \{q \in Q \mid q = q_1 q_2 \ldots q_k \text{ with } q_k \in F_A \text{ or } k > 1 \text{ and } q_{k-1} \in F_A\}.$$

Observe, that a *reachable* final state $q = q_1 q_2 \ldots q_k$ with $q_{k-1} \in F_A$ must fulfill $q_k = q_{0,A}$ by construction. Moreover, if $q = q_1 q_2 \ldots q_k$ is a *reachable* final state with $q_k \in F_A \setminus \{q_{0,A}\}$, then we must have $q_{0,A} \in \{q_1, q_2, \ldots, q_{k-1}\}$. The language accepted by C is $L(C) = L^{!+}$. Because $L^{!*} = L^{!+} \cup \{\lambda\}$, a DFA for $L^{!*}$ can be obtained from C by simply introducing an additional *accepting* copy of the initial state of C (unless $q_{0,A} \in F_A$ and thus $L(C) = L^{!+} = L^{!*}$).

In the forthcoming sections we consider the descriptional complexity of both operations, when the regular languages are given by DFAs. The above presented constructions show an asymptotic upper bound of $O(n \cdot (m+1))$ for the cut operation, and an asymptotic upper bound of $O(n!)$ for the iterated cut operation, if A and B are DFAs with n and m states, respectively.

3   The Descriptional Complexity of the Cut Operation

In this section we prove a tight bound for a DFA accepting the cut of two languages, when these languages are represented by an n- and m-state DFA, respectively. This exact tight bound is $(n-1) \cdot m + n$, which is witnessed by automata using binary input alphabets. Then we consider the special case of unary languages. Here we have to do a detailed analysis of the structure of unary DFAs in order to prove a tight bound of $\max(2n-1, n+m-2)$ on the number of states for unary DFAs. Notice that for $m \le n$ this bound only depends on the first automaton, but not on the second.

First we consider a few special cases. Let Σ be the input alphabet of the DFA A. If all states in A are accepting, then $L(A) = \Sigma^*$, and if A does not have an accepting state at all, then $L(A) = \emptyset$. Thus in both cases, the cut of $L = L(A)$ with any other language L', i.e., $L\,!\,L'$, is empty or equal to Σ^* (the latter being the case if $L = \Sigma^*$ and $\lambda \in L'$). Thus, in each of these cases the resulting language can be described by a DFA with single state only. In general we obtain the following result.

Theorem 1. *Let A be an n-state and B an m-state deterministic finite automaton. Then $(n-1) \cdot m + n$ states are sufficient and necessary in the worst case for any deterministic finite automaton accepting the language $L(A)\,!\,L(B)$. The lower bound even holds for automata with binary input alphabet.*

Proof. Let $A = (Q_A, \Sigma, \delta_A, q_{0,A}, F_A)$ and $B = (Q_B, \Sigma, \delta_B, q_{0,B}, F_B)$ be the n- and m-state DFAs, respectively. Applying the previously described construction for the cut gives a DFA C that accepts $L(A)\,!\,L(B)$ with a state set consisting of stack states of height at most two. A careful inspection of this construction reveals that some stack states are not reachable, since they do not have any incoming transitions. (i) Among the stack states of height one only those that are *not* final states in A are possibly reachable. (ii) For the stack states of height two we consider two sub-cases, namely if the first element of the sequence is in F_A, then the second one is always the initial state of B, and if the first element of the sequence is in $Q_A \setminus F_A$, the second one may be an arbitrary element of Q_B. Thus, the state set Q_C of C can be restricted to contain stack states from

$$Q_C = \{\, p \mid p \in Q_A \setminus F_A \,\} \cup F_A\{q_{0,B}\} \cup \{\, pq \mid p \in (Q_A \setminus F_A) \text{ and } q \in Q_B \,\}$$

without changing the accepted language. Thus, we have at most

$$n - |F_A| + |F_A| + (n - |F_A|) \cdot m = (n - |F_A|) \cdot m + n$$

states, which is maximal if A has only a single accepting state, leading to an upper bound of $(n-1) \cdot m + n$.

Next we show that this upper bound can be reached. To this end we define the n-state DFA $A = (Q_A, \{a, b\}, \delta_A, q_{0,A}, F_A)$, where $Q_A = \{0, 1, \ldots, n-1\}$, $q_{0,A} = 0$, $F_A = \{n-1\}$, and

$$\delta_A(i, a) = i + 1 \pmod{n} \quad \text{and} \quad \delta_A(i, b) = i, \quad \text{for } 0 \le i \le n-1.$$

Moreover, we define the m-state DFA $B = (Q_B, \{a, b\}, \delta_B, q_{0,B}, F_B)$, where $Q_B = \{0, 1, \ldots, m-1\}$, $q_{0,B} = 0$, $F_B = \{0\}$, and

$$\delta_B(i, a) = i, \quad \text{for } 0 \le i \le m-1 \quad \text{and} \quad \delta_B(i, b) = i + 1 \pmod{m}$$

Both finite automata are depicted in Fig. 1. Again, let C be the DFA constructed from A and B by applying the construction for the cut, where the state set Q_C is restricted according to our previous considerations. Moreover, let $q_{0,C}$ be the initial state of C. To prove that C needs exactly the claimed number of states, it can be shown that all states in Q_C are reachable and pairwise distinguishable. Due to space constraints the proof is omitted.   □

As we have seen in the previous proof, the upper bound on the number of states for the cut of two DFAs implicitly depends on the number of accepting states of the first automaton. It is easy to generalize the above argument to lead to an even more precise tight upper and lower bound of $(n - f) \cdot m + n$ states, where f with $1 \le f < n$ refers to the number of accepting states of the "left" automaton A. For the lower bound proof one simply has to alter the definition of the automaton A by setting its accepting states to $F_A = \{n-f, \ldots, n-2, n-1\}$. The details are left to the reader.

In the remainder of this section we briefly mention (without proofs) the descriptional complexity of the cut operation on unary languages, that is, languages over a single letter alphabet. Deterministic finite automata for unary

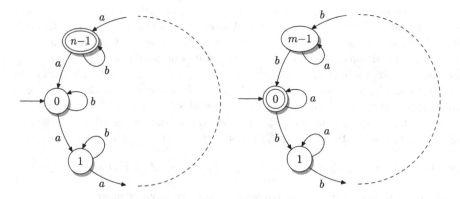

Fig. 1. The binary DFAs A (left) and B (right) with n and m states, respectively, that witness the state complexity lower bound for the cut operation.

languages obey a very simple structure: a (possibly empty) initial chain, followed by a cycle.

Theorem 2. *Let A be an n-state and B an m-state deterministic finite automaton accepting a unary language. Then $g(n, m)$ states, where*

$$g(n, m) = \begin{cases} 1 & \text{if } n = 1, \\ n & \text{if } n \geq 2 \text{ and } m = 1, \\ 2n - 1 & \text{if } n, m \geq 2 \text{ and } m \leq n, \\ n + m - 2 & \text{if } n, m \geq 2 \text{ and } m > n, \end{cases}$$

are sufficient and necessary in the worst case for any deterministic finite automaton accepting the language $L(A) \, ! \, L(B)$. □

Similarly to the non-unary case, the state complexity of the cut on unary DFAs also depends on the number of accepting states of the "left" automaton A. If $L(A)$ is infinite, and automaton A has f accepting states, then we obtain a tight bound of $2n - f$ states for a DFA that accepts the language $L(A) \, ! \, L(B)$. However, if $L(A)$ is finite then the bound stays $n + m - 2$, regardless of the number of accepting states.

Sometimes, when studying descriptional complexity of unary regular languages, one does not simply count the number of states of a DFA, but rather distinguishes between the length of the initial chain and the length of the cycle. So instead of asking for the number $g(n, m)$ of states of a DFA for accepting the cut of the languages described by n-state and m-state DFAs, one could also study the following: given unary DFAs A_1 and A_2, with t_1 (t_2, respectively) states in the initial chain and k_1 (k_2, respectively) states in the cycle, determine bounds (as functions in t_1, t_2, k_1, k_2) for the number t of states in the initial chain and the number k of states in the cycle of a DFA for the language $L(A_1) \, ! \, L(A_2)$. Since the results on tight bounds for t and k branch out into many different sub-cases, we will not go into details here, but rather summarize our results in Table 1.

4 The Descriptional Complexity of the Iterated-Cut Operation

In this section we turn our attention to the state complexity of the iterated cut operation on DFAs. In order to properly state our result we need some more notation. A *generalized hypergeometric function* [4] is a power series in x with $r + s$ parameters, and it is defined as follows in terms of rising factorial powers:

$$\mathsf{F}\left(\begin{array}{c} a_1, a_2, \ldots, a_r \\ b_1, b_2, \ldots, b_s \end{array} \middle| x\right) = \sum_{\ell \geq 0} \frac{a_1^{\overline{\ell}} a_2^{\overline{\ell}} \ldots a_r^{\overline{\ell}}}{b_1^{\overline{\ell}} b_2^{\overline{\ell}} \ldots b_s^{\overline{\ell}}} \cdot \frac{x^\ell}{\ell!}.$$

Here the *rising factorial* is defined as $x^{\overline{\ell}} = x(x + 1) \cdots (x + (\ell - 1))$ and the *falling factorial* by $x^{\underline{\ell}} = x(x - 1) \cdots (x - (\ell - 1))$. By convention $x^{\overline{0}} = x^{\underline{0}} = 1$.

Table 1. Tight bounds for the length t of the initial chain and the length k of the cycle of a DFA for the language $L(A_1)\,!\,L(A_2)$, where A_i has an initial chain of length t_i and a cycle of length k_i, for $i = 1, 2$. The table is ordered by ascending values of k_1. If a tuple (t_1, k_2, t_2, k_2) matches multiple lines, the additional condition column has to be checked.

t_1	k_1	t_2	k_2	condition	t	k
0	1	≥ 0	≥ 1		0	1
≥ 1	1	≥ 0	≥ 1	accepting state in cycle 1	t_1	1
≥ 1	1	≥ 0	≥ 1	no accepting state in cycle 1	$t_1 + t_2 - 1$	k_2
≥ 0	≥ 2	0	1		$t_1 + k_1 - 1$	1
≥ 0	≥ 2	≥ 1	1		$t_1 + k_1 - 1$	k_1
≥ 0	2	0	≥ 2		$t_1 + k_1 - 1$	k_1
≥ 0	2	1	2		$t_1 + k_1 - 1$	1
≥ 0	2	1	≥ 3		$t_1 + k_1 - 1$	k_1
≥ 0	2	≥ 2	≥ 2		$t_1 + k_1 - 1$	k_1
≥ 0	≥ 3	≥ 0	≥ 2	$k_1 \leq t_2 + k_2$	$t_1 + k_1 - 1$	k_1
≥ 0	≥ 3	0	≥ 2	$k_1 > k_2$ and $k_2 \bmod k_1 > 0$	$t_1 + k_1 - 1$	k_1
≥ 0	≥ 3	0	≥ 2	$k_1 > k_2$, and $k_1 = k \cdot k_2$, and 1 accepting in k_1-loop	$t_1 + k_1 - 1$	k_2
≥ 0	≥ 3	0	≥ 2	$k_1 > k_2$, and $k_1 = k \cdot k_2$, and ≥ 2 accepting in k_1-loop	$t_1 + k_1 - 2$	k_1
≥ 0	≥ 3	≥ 1	≥ 2	$k_1 > t_2 + k_2$	$t_1 + k_1 - 1$	k_1

Then our result on the number of states that are sufficient and necessary to accept the iterated cut $L^{!+}$ of a single language accepted by an n-state DFA reads as follows—a corresponding result for $L^{!*}$ will be given later.

Theorem 3. *Let A be a deterministic finite automaton with n states. Then*

$$(n+1) \cdot \mathsf{F}\left(\begin{array}{c} 1, n+2, -n+2 \\ n+1 \end{array} \middle| -1 \right)$$

states are sufficient and necessary in the worst case for a deterministic finite automaton to accept the language $L(A)^{!+}$. The lower bound even holds for automata with ternary input alphabet.

Before we prove this theorem by the upcoming two lemmata, we first show that the following combinatorial identity holds.

Theorem 4. *For natural numbers n with $n \geq 2$ we have the identity*

$$(n+1) \cdot \mathsf{F}\left(\begin{array}{c} 1, n+2, -n+2 \\ n+1 \end{array} \middle| -1 \right) = \sum_{\ell=0}^{n-2} (n+\ell+1) \cdot (n-2)^{\underline{\ell}}. \tag{1}$$

Proof. The proof outline follows the presentation on hypergeometric functions given in [4]. Note that the sum on the right hand-side can be changed to sum up for all ℓ with $\ell \geq 0$, because for $\ell > (n-2)$ the falling factorials $(n-2)^{\underline{\ell}}$ are always zero, and thus these terms do not contribute anything to the sum.

Now let the notation of the series be $\sum_{\ell \geq 0} t_\ell$ with $t_0 \neq 0$. If the term ratio $t_{\ell+1}/t_\ell$ is a rational function in ℓ, that is, a quotient of polynomials in ℓ of the form

$$\frac{(\ell + a_1)(\ell + a_2)\ldots(\ell + a_r)}{(\ell + b_1)(\ell + b_2)\ldots(\ell + b_s)} \cdot \frac{x}{(\ell + 1)}$$

then we can use the ansatz

$$\sum_{\ell \geq 0} t_\ell = t_0 \cdot \mathsf{F}\left(\begin{array}{c} a_1, a_2, \ldots, a_r \\ b_1, b_2, \ldots, b_s \end{array} \middle| x \right).$$

As $t_\ell = (n + \ell + 1)(n - 2)^\ell$, the first term of our sum is $t_0 = (n + 1)$, and the other terms have the ratios given by

$$\frac{t_{\ell+1}}{t_\ell} = \frac{(n + \ell + 2)(n - 2)^{\ell+1}}{(n + \ell + 1)(n - 2)^\ell} = \frac{(n + \ell + 2)(n - 2 - \ell)}{(n + \ell + 1)},$$

which are rational functions of ℓ. Rearranging the terms and introducing the required factor $(\ell + 1)$ in the denominator results in

$$\frac{t_{\ell+1}}{t_\ell} = \frac{(\ell + 1)(\ell + n + 2)(\ell - n + 2)}{(\ell + n + 1)} \cdot \frac{(-1)}{(\ell + 1)},$$

where we can read off the result: the given sum is equal to

$$(n + 1) \cdot \mathsf{F}\left(\begin{array}{c} 1, n + 2, -n + 2 \\ n + 1 \end{array} \middle| -1 \right),$$

which proves the stated result.   □

The first few values of the hypergeometric function in Equation (1) starting with $n = 1$ are $1, 3, 9, 31, 129, 651, 3913, 27399, 219201, 1972819, 19728201, \ldots$. In the On-Line Encyclopedia of Integer Sequences (OEIS)—see www.oeis.org—this matches the sequence A111063. A detailed analysis of the behaviour of this sequence is given after the following two lemmata that prove Theorem 3.

Lemma 5. *Let A be deterministic finite automaton with n states. Then*

$$\sum_{\ell=0}^{n-2} (n + \ell + 1) \cdot (n - 2)^\ell$$

states are sufficient for a deterministic finite automaton to accept $L(A)^{!+}$.

Proof. The upper bound can be seen as follows. Let $A = (Q, \Sigma, \delta, q_{0,A}, F)$ be a DFA, and C be the DFA as constructed in Sect. 2 for accepting $L(A)^{!+}$. By construction the state set of C is $\pi(Q^+)$, consisting of stacks of states from Q. Every such stack of height $\ell \geq 1$ is of one of the following forms, called *types*—recall that by the definition of π, all elements in a stack state are pairwise distinct:

Type 1: $q = q_1 q_2 \ldots q_\ell$, with $q_1, q_2 \ldots, q_\ell \in Q \setminus F$, or

Type 2: $q = q_1 q_2 \ldots q_{\ell-1} q_{0,A}$, with $q_1, q_2, \ldots, q_{\ell-1} \in Q \setminus F$, and $q_{0,A} \in F$, or

Type 3: $q = q_1 q_2 \ldots q_{\ell-1} q_\ell$, with $q_1, q_2, \ldots, q_{\ell-1} \in Q \setminus F$, and $q_\ell \in F$, and $q_{0,A} \in \{q_1, q_2, \ldots q_{\ell-1}\}$, or

Type 4: $q = q_1 q_2 \ldots q_{\ell-2} q_{\ell-1} q_{0,A}$ (and therefore $q_{0,A} \notin \{q_1, q_2, \ldots q_{\ell-1}\}$), with $q_1, q_2, \ldots, q_{\ell-2} \in Q \setminus F$ and $q_{\ell-1} \in F$.

Let us count the number of states of the different types. Clearly, the number of different stacks of type 1 is $\sum_{\ell=1}^{n-|F|} (n-|F|)^\ell$, and the number of type 2 stacks is $\sum_{\ell=1}^{n-|F|+1} (n-|F|)^{\ell-1}$. To build a stack of type 3 we choose $\ell-2$ non-accepting, non-initial states, permute them, then shuffle $q_{0,A}$ somewhere into these states, and put an accepting state on top. This gives $\sum_{\ell=2}^{n-|F|+1} (n-|F|-1)^{\ell-2} \cdot (\ell-1) \cdot |F|$ different stacks. Finally, to count the number of stacks of type 4, we distinguish between the two cases $q_{0,A} \in F$ and $q_{0,A} \notin F$. In the former case, a stack is built by choosing and permuting $\ell-2$ non-accepting states, then putting an accepting, non-initial state and state $q_{0,A}$ on top—this gives $\sum_{\ell=2}^{n-|F|+2} (n-|F|)^{\ell-2} \cdot (|F|-1)$ different stacks of type 4. Similarly, for the case where $q_{0,A} \notin F$ we choose and permute $\ell-2$ non-accepting, non-initial states, put an accepting state and then state $q_{0,A}$ on top, which gives $\sum_{\ell=2}^{n-|F|+1} (n-|F|-1)^{\ell-2} \cdot |F|$ stacks.

The bound in the statement of the lemma will result from the case where $|F| = 1$ and $q_{0,A} \notin F$. To see that this case indeed yields an upper bound for all the cases, we first argue that the overall number of different possible stacks increases when the number of accepting states decreases.

Given a deterministic finite automaton $A = (Q, \Sigma, \delta, q_{0,A}, F)$ with $|F| \geq 2$, we construct an automaton $B = (Q, \Sigma, \delta, q_{0,A}, F')$ such that

$$F' = \begin{cases} F \setminus \{q_{0,A}\} & \text{if } q_{0,A} \in F, \\ F \setminus \{q_f\} & \text{for some } q_f \in F \text{ if } q_{0,A} \notin F. \end{cases}$$

Denote by $\mathcal{S}(Q, F)$ the set of stacks that can be built by applying the automaton construction for $L^{!+}$ to automaton A, and by $\mathcal{S}(Q, F')$ those that can be built by applying the construction to B. We want to show $|\mathcal{S}(Q, F)| \leq |\mathcal{S}(Q, F')|$. In fact, treating the stacks as words over alphabet Q, we show the inclusion $\mathcal{S}(Q, F) \subseteq \mathcal{S}(Q, F')$. Clearly, every type 1 stack in $\mathcal{S}(Q, F)$ also appears as type 1 stack in $\mathcal{S}(Q, F')$, since $Q \setminus F$ is a subset of $Q \setminus F'$. Now assume we have a type 2 stack $q_1 q_2 \ldots q_{\ell-1} q_{0,A} \in \mathcal{S}(Q, F)$, which means that $q_{0,A} \in F$. Then this stack $q_1 q_2 \ldots q_{\ell-1} q_{0,A}$ is a type 1 stack in $\mathcal{S}(Q, F')$. For stacks $q_1 q_2 \ldots q_{\ell-1} q_\ell \in \mathcal{S}(Q, F)$ of type 3 we have $q_\ell \in F$ and $q_{0,A} \notin F$. Here we distinguish between the two cases $q_\ell \in F'$ and $q_\ell \notin F'$. In the former case, the stack $q_1 q_2 \ldots q_{\ell-1} q_\ell$ also appears as type 3 stack in $\mathcal{S}(Q, F')$, and in case $q_\ell \notin F'$ this stack is of type 1 in $\mathcal{S}(Q, F')$. The argumentation for stacks $q_1 q_2 \ldots q_{\ell-2} q_{\ell-1} q_{0,A} \in \mathcal{S}(Q, F)$ of type 4, where $q_{\ell-1} \in F$ is similar: if $q_{\ell-1} \in F'$, then $q_1 q_2 \ldots q_{\ell-2} q_{\ell-1} q_{0,A}$ is also a type 4 stack in $\mathcal{S}(Q, F')$, and if $q_{\ell-1} \notin F'$, then it appears in $\mathcal{S}(Q, F')$ as a stack of type 1. We have shown $\mathcal{S}(Q, F) \subseteq \mathcal{S}(Q, F')$, so the smaller the set of accepting states, the larger is the number of possible stacks. Therefore, the case where $|F| = 1$ forms an upper bound—we ignore $|F| = 0$ since the accepted language would be empty. From [1] we already know that languages accepted by DFAs where the initial state is the sole accepting state are closed under iterated cut. Thus, for the upper bound we choose the case $|F| = 1$ and $q_{0,A} \notin F$. Using the sums from above we obtain

$$\sum_{\ell=1}^{n-1}(n-1)^{\ell} + \sum_{\ell=2}^{n}(n-2)^{\ell-2} \cdot (\ell - 1) + \sum_{\ell=2}^{n}(n-2)^{\ell-2}$$

$$= \sum_{\ell=1}^{n-1}(n-1)\cdot(n-2)^{\ell-1} + (n-2)^{\ell-1} \cdot (\ell + 1)$$

$$= \sum_{\ell=0}^{n-2}(n+\ell+1)\cdot(n-2)^{\ell}$$

as an upper bound for the number of states of a DFA for the language $L(A)^{!+}$. Notice that we ignore type 2 stacks and choose the sum for the second case of type 4 stacks, since we have $q_{0,A} \notin F$. □

The next lemma provides a matching lower bound, and thus concludes the proof of Theorem 3.

Lemma 6. *For every $n \geq 4$, there exists a deterministic finite automaton A with n states, such that the number of states of the minimal deterministic finite automaton for the language $L(A)^{!+}$ is*

$$\sum_{\ell=0}^{n-2}(n+\ell+1)\cdot(n-2)^{\ell}.$$

Proof. Let $n \geq 4$ and $A = (Q, \Sigma, \delta, 0, F)$ with input alphabet $\Sigma = \{a, b, c\}$, state set $Q = \{0, 1, \ldots, n-1\}$ and accepting states $F = \{n-1\}$ be the DFA depicted in Fig. 2. The transition function δ is defined as follows:

$$\delta(q, a) = \begin{cases} q+1 & \text{if } 0 \leq q \leq n-3, \\ 0 & \text{if } q = n-2, \\ n-1 & \text{if } q = n-1, \end{cases}$$

$$\delta(q, b) = \begin{cases} 1 & \text{if } q = 0, \\ 0 & \text{if } q \in \{1, n-1\}, \\ q & \text{if } 2 \leq q \leq n-2, \end{cases}$$

and

$$\delta(q, c) = \begin{cases} n-1 & \text{if } q = 0, \\ q & \text{if } 1 \leq q \leq n-2, \\ 0 & \text{if } q = n-1. \end{cases}$$

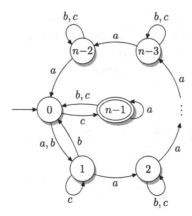

Fig. 2. The n-state DFA A for witnessing the state complexity lower bound for the iterated cut operation.

First notice, that the two mappings $q \mapsto \delta(q, a)$ and $q \mapsto \delta(q, b)$ generate all permutations on the set $Q \setminus F$—see, e.g., [7]. Let $C = (Q_C, \Sigma, \delta_C, q_{0,C}, F_C)$ be the DFA constructed from A by applying the construction for $L^{!+}$. To prove that C needs exactly the claimed number of states, we show that all states, or stacks, of types 1, 3, and 4 from the proof of Lemma 5 are reachable and pairwise distinguishable—note that stacks of type 2 do not appear in C.

Reachability of stacks of type 1 is easy to see: using permutations on the set $Q \setminus F$, which can be realized by reading appropriate words over $\{a, b\}$, every type 1 stack can be transformed into every other type 1 stack of the same size. Moreover, from a type 1 stack of the form $q_1 q_2 \ldots q_{\ell-1} 0$ of size $\ell \leq n - 2$, with $1 \notin \{q_1, q_2, \ldots, q_{\ell-1}\}$, we can reach a stack $q_1 q_2 \ldots q_{\ell-1} 1 0$ of type 1 with size $\ell+1$ by reading cb. Now every type 4 stack $q_1 q_2 \ldots q_{\ell-2} (n-1) 0$ can be reached from the type 1 stack $q_1 q_2 \ldots q_{\ell-2} 0$ by reading c. From type 4 stacks we can reach type 3 stacks as follows. If the wanted stack has the initial state 0 directly below state $(n-1)$, that is, if it has the form $q_1 q_2 \ldots q_{\ell-2} 0 (n-1)$, then we can reach it from $q_1 q_2 \ldots q_{\ell-2} (n-1) 0$ by simply reading c. If the element 0 is not directly below element $(n-1)$, that is, if we want to obtain a stack of the form

$$q_1 q_2 \ldots q_{i-1} 0 q_i q_{i+1} \ldots q_{\ell-2} (n-1),$$

with $i \leq \ell - 2$, then we can reach it from the type 3 stack

$$p_1 p_2 \ldots p_{i-1} r p_i p_{i+1} \ldots p_{\ell-2} (n-1),$$

by reading $a^{q_{\ell-2}}$, where $p_j = (q_j - q_{\ell-2}) \bmod (n-2)$ for $1 \leq j \leq \ell - 2$, and $r = (-q_{\ell-2}) \bmod (n-2)$. Notice that $p_{\ell-2} = 0$, so we already know how to reach the latter stack from a type 4 stack.

It remains to prove that every pair of states, or stacks, s_1 and s_2 of C can be distinguished by some input word. Clearly we only have to consider the cases where either both stacks contain the accepting state $n-1$, or none of them does.

We start with the case where $n - 1$ does not appear in the stacks. If there is some element q_i that appears in one of the two stacks but not in the other, then we can use a permutation that interchanges q_i and 0, and leaves the other elements stable—if $q_i = 0$, we just take the identity permutation. Now one stack contains element 0 and the other does not, so we can distinguish between those two by reading c. Now assume that both stacks contain the same elements, but differ in their ordering. If the two stacks already differ in their bottom elements, that is, if $s_1 = q_1 t_1$ and $s_2 = q_2 t_2$ with $q_1 \neq q_2$ and appropriate sequences t_1 and t_2, then we use a permutation to interchange q_1 and 0, and obtain stacks $0\,t_1'$ and $q_2' t_2'$, with $q_2' \neq 0$. With a permutation for interchanging q_2' and 2, we obtain stacks $0\,t_1''$ and $2\,t_2''$. These are distinguished by reading cb: the second stack yields a stack containing the element 2, but the stack $0\,t_1''$ yields $0\,1$, which does not contain this element, and we have seen above how to distinguish the stacks in this case.

Next we show that we can also distinguish between stacks s_1 and s_2 that both contain the accepting state $n - 1$. Let us consider the lowest (leftmost) position in which the two stacks differ. We have $s_1 = s_0\,q_1\,t_1$ and $s_2 = s_0\,q_2\,t_2$, for appropriate sequences s_0, t_1, t_2 and elements $q_1 \neq q_2$. If at least one of the elements q_1 and q_2 is from the set $\{2, 3, \ldots, n-2\}$, then the stacks obtained from s_1 and s_2 after reading b are still different because every state $q \in \{2, 3, \ldots, n-2\}$ loops on input b and no other b-transitions lead to q. Now there are only three cases remaining, namely $\{q_1, q_2\} = \{0, 1\}$, $\{q_1, q_2\} = \{0, n - 1\}$, and $\{q_1, q_2\} = \{1, n - 1\}$. In the first two cases, where one of the elements q_1 and q_2 is 0, we can also read input b to get rid of element $n-1$, and again we obtain two different stacks because the transition from 0 to 1 is the only b-transition that leads to state 1. For the remaining case, we may assume $q_1 = 1$ and $q_2 = n - 1$, so we have stacks

$$s_1 = s_0\,1\,t_1 \qquad \text{and} \qquad s_2 = s_0\,(n - 1)\,t_2,$$

for appropriate sequences s_0, t_1, and t_2, where in particular element 1 does not appear in s_0. Now we read the input word ab. First, after reading a we have stacks

$$s_1' = s_0'\,2\,t_1' \qquad \text{and} \qquad s_2' = s_0'\,(n - 1)\,0,$$

where element 2 does not appear in s_0', and thus, not in s_2'. Now by reading the b symbol, stack s_1' yields a stack of the form $s_1'' = s_0''\,2\,t_1''$ that contains element 2, while stack s_2' results in stack $s_2'' = s_0''\,0$ or $s_2'' = s_0''\,0\,1$, depending on whether element 1 appears in s_0''. In either case the stacks s_1'' and s_2'' are different and none of them contains element $n - 1$, so they can be distinguished as described earlier. This concludes our proof. $\square$

Recall that the DFA for $L^!{}^+$ can be turned into a DFA for $L^!{}^*$ by adding a new accepting initial state. Therefore, the upper bound for the state complexity for $L^!{}^*$ is by one larger than the bound for $L^!{}^+$. In fact, one can see as follows that this bound is also tight by using the same witness automaton as in the previous proof. Let $q'_{0,C}$ be the new accepting initial state; it has the same outgoing transitions as state $q_{0,C} = 0$. We only need to distinguish state $q'_{0,C}$ from all other accepting states of C, which are the stack states that contain element $(n-1)$. This can simply be done by reading letters a: the successor of the accepting stack state also contains element $(n-1)$, and thus is accepting, but the successor of state 0, and thus also of $q'_{0,C}$, is the non-accepting state 1. Therefore we obtain the following result on the state complexity of the iterated cut $L^!{}^*$.

Theorem 7. *Let A be a deterministic finite automaton with n states. Then*

$$1 + (n+1) \cdot \mathsf{F}\left(\begin{array}{c} 1, n+2, -n+2 \\ n+1 \end{array}\middle| -1\right)$$

states are sufficient and necessary in the worst case for a deterministic finite automaton to accept the language $L(A)^!{}^*$. The lower bound even holds for automata with ternary input alphabet. □

As in the case of the (non-iterated) cut operation, one can also derive a more precise bound for the state complexity of the iterated cut operation, which depends on the number f of accepting states of the automaton. From the upper bound analysis in the proof of Lemma 5, a bound of

$$\sum_{\ell=0}^{n-f-1} (n-f-1)^{\ell} \cdot (n + f \cdot (\ell+1))$$

states can be derived for the case where the initial state $q_{0,A}$ is not accepting. In fact, by some calculations and combinatorial argumentation, one can show that the case, where the initial state *is* accepting, yields the same upper bound. By adapting the automaton from the lower bound proof of Lemma 6, one can obtain witness automata for the more precise bound as follows. Instead of a single accepting state $n-1$ we choose f accepting states $n-1, n-2, \ldots, n-f$, that are connected to a cycle of $n-f$ non-accepting states in a similar fashion as shown in Fig. 2: for each accepting state $n-i$, for $1 \le i \le f$, we use another input symbol c_i that switches between states $n-i$ and 0; on input symbol b the states $n-i$ also go to state 0.

Now let us come to the asymptotics of the bounds stated in Theorems 3 and 7 in order to get a better feeling for their size. This can be done by first proving the following upper and lower bounds.

Theorem 8. *The following lower and upper bounds apply:*

$$2 \cdot (n+2) \cdot (n-2)! \le \sum_{k=0}^{n-2}(n+k+1)(n-2)^{\underline{k}} \le e \cdot (2n-1) \cdot (n-2)!$$ □

The upper and the lower bound are quite close, since the former is asymptotically only a factor of e away from the latter. This allows us to show that the bounds provided in Theorems 3 and 7 in the exact number of states asymptotically behave like $(n-1)!$.

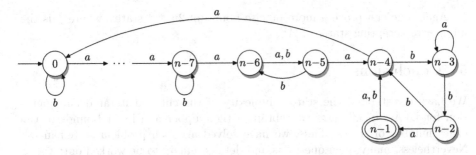

Fig. 3. A DFA A with binary input alphabet $\Sigma = \{a, b\}$, where the minimal DFA for $L(A)^{!+}$ needs at least $\sum_{\ell=1}^{n-3}(n-3)^{\underline{\ell}} > (n-3)!$ states.

Theorem 9. *Let A be a deterministic finite automaton with n states. Then $\Theta((n-1)!)$ states are sufficient and necessary in the worst case for a deterministic finite automaton to accept the language $L(A)^{!+}$. The same holds for a deterministic finite automaton for $L(A)^{!*}$.* $\square$

Notice that the witness automaton from the proof of Lemma 6 uses a ternary input alphabet. Theorem 10 will give a tight bound of $2n - 1$ for the state complexity of the iterated cut operation on DFAs with unary input alphabet. So it remains to study the exact bound for automata using a binary alphabet. Here we do not have a tight bound yet. However, we know that already a binary alphabet is enough to provide a huge blow-up in the number of states. Consider the binary n-state DFA A depicted in Fig. 3. One can show that the minimal DFA for the language $L(A)^{!+}$ needs more than $\sum_{\ell=1}^{n-3}(n-3)^{\underline{\ell}} > (n-3)!$ states. The basic idea to prove this is that all stack states with elements from the set $\{0, 1, \ldots, n-4\}$ can be reached: permutations on this set can be obtained with the help of the input words a (cycling through the set) and b^3 (switching elements $n-5$ and $n-6$). New elements can be obtained by reading b^2ab^4 from a stack of the form $s = s'(n-4)$, where $n-5$ does not appear in s'.

Now we come to our result on the state complexity of the iterated cut operation on unary languages. The upcoming theorem only considers input automata with at least three states. The reader may convince him- or herself that every minimal two-state DFA A (there are only four possibilities) yields as iterated cut $L(A)^{!*}$ a language that can also be accepted by a DFA with at most two states. Moreover, the iterated cut languages of the single-state languages $\emptyset$ and $\{a\}^*$ are $\{\lambda\}$ and $\{a^*\}$ respectively, and thus are accepted by a two-state, respectively, single-state DFA.

We now state the already mentioned theorem providing the tight bound for the iterated cut in the case of a unary alphabet.

Theorem 10. *Let A be a deterministic finite automaton with $n \geq 3$ states that accepts a unary language. Then $2n - 1$ states are sufficient and necessary in the worst case for a deterministic finite automaton to accept the language $L(A)^{!+}$. The same holds for a deterministic finite automaton for $L(A)^{!*}$.* $\square$

Again one can prove a more precise bound of $2n - f$ states, where f is the number of accepting states of A.

5 Conclusions

We have investigated the state complexity of the cut and iterated cut operation for DFAs. In all cases, we obtained tight upper and lower bounds in the exact number of states. Thus, we have solved an open problem stated in [1]. Nevertheless, many open questions and details remain to be worked out:

- Consider the upper and lower bounds for nondeterministic finite automata (NFAs) on cut operations. Can we do better than first determinizing the involved devices and then performing the cut or iterated cut construction for DFAs? Note that for cuts one does not need to determinize the second automaton B in order to construct the (then nondeterministic) automaton for $L(A) ! L(B)$ as in Sect. 2.
- The complexity of decision problems related to the cut and iterated cut operation on finite automata, in particular, on DFAs. An example of such a problem is the following: given a finite automaton A, is $L(A)^{!*} = L(A)^*$?
- Succinctness of cut expressions (these are regular expressions that also use the cut operation) compared to DFAs and NFAs were discussed in [1]. There exponential lower bounds for both types of finite state devices were obtained. So what about the succinctness of iterated cut expressions (regular expressions that also use iterated cut) compared to finite automata?

Acknowledgments. Thanks to Rogério Reis for his help doing and verifying some calculations with the computer algebra system MAPLE^{TM}.

References

1. Berglund, M., Björklund, H., Drewes, F., van der Merwe, B., Watson, B.: Cuts in regular expressions. In: Béal, M.-P., Carton, O. (eds.) DLT 2013. LNCS, vol. 7907, pp. 70–81. Springer, Heidelberg (2013)
2. Berglund, M., Drewes, F., van der Merwe, B.: Analyzing catastrophic backtracking behavior in practical regular expressions matching. In: Ésik, Z., Fülöp, Z. (eds.) Proceedings of the 14th International Conference on Automata and Formal Languages. EPTCS, vol. 151, pp. 109–123, Szeged, Hungary (2014)
3. Clocksin, W.F., Mellish, C.S.: Programming in Prolog. Springer, Heidelberg (1981)
4. Graham, R.L., Knuth, D.E., Patashnik, O.: Concrete Mathematics: A foundation for Computer Science. Addison-Wesley, Boston (1994)
5. Harrison, M.A.: Introduction to Formal Language Theory. Addison-Wesley, Boston (1978)
6. Kirrage, J., Rathnayake, A., Thielecke, H.: Static analysis for regular expression denial-of-service attacks. In: Lopez, J., Huang, X., Sandhu, R. (eds.) NSS 2013. LNCS, vol. 7873, pp. 135–148. Springer, Heidelberg (2013)
7. Piccard, S.: Sur les bases du groupe symétrique et les couples de substitutions qui engendrent un groupe régulier. Librairie Vuibert, Paris (1946)

Non-isometric Contextual Array Grammars with Regular Control and Local Selectors

Henning Fernau[1]([⊠]), Rudolf Freund[2], Rani Siromoney[3],
and K.G. Subramanian[4]

[1] Universität Trier, FB 4 – Abteilung Informatikwissenschaften,
54296 Trier, Germany
fernau@uni-trier.de
[2] Institut für Computersprachen, TU Wien, 1040 Vienna, Austria
rudi@emcc.at
[3] Chennai Mathematical Institute, Kelambakkam 603103, India
siromoney@cmi.ac.in
[4] Department of Mathematics and Computer Science, Faculty of Science,
Liverpool Hope University, Liverpool L16 9JD, Great Britain
kgsmani1948@yahoo.com

Abstract. We consider the external variant of non-isometric d-dimensional contextual array grammars with regular control together with local selectors allowing for controlling how d-dimensional arrays are evolving by adjoining rectangular $(d-1)$-dimensional arrays. In the 1-dimensional case, the computational power of these non-isometric contextual array grammars with regular control and local selectors equals the computational power of isometric contextual array grammars with regular control. The string images of the langauges of 1-dimensional arrays generated by these contextual array grammars exactly yield the linear languages. In the more-dimensional case, non-isometric d-dimensional contextual array grammars with regular control and local selectors can simulate the computations of $(d-1)$-dimensional array grammars or Turing machines. Hence, for example, the emptiness problem for non-isometric d-dimensional contextual array grammars with regular control and local selectors for $d > 1$ is undecidable.

1 Introduction

Contextual string grammars were introduced by Solomon Marcus [11] with motivations arising from descriptive linguistics. A contextual string grammar consists of a finite set of strings (*axioms*) and a finite set of productions, which are pairs (s, c) where s is a string, the *selector,* and c is the *context,* i.e., a pair of strings, $c = (u, v)$, over the alphabet under consideration. Starting from an axiom, contexts iteratively are added as indicated by the productions, which yields new strings. In contrast to usual sequential string grammars in the Chomsky hierarchy (e.g., see [18,19]), these contextual string grammars are pure grammars where new strings are not obtained by rewriting, but by adjoining strings. If the contexts are adjoined at both ends of a string; this case is named *external* (see [11]);

© Springer International Publishing Switzerland 2015
J. Durand-Lose and B. Nagy (Eds.): MCU 2015, LNCS 9288, pp. 61–78, 2015.
DOI: 10.1007/978-3-319-23111-2_5

in the *interior* case (see [13]), one occurrence of a selector is bracketed with the associated contexts. Some more variants of contextual grammars have been introduced and investigated, e.g., see [1,14] for surveys on the area.

In the area of two-dimensional picture languages, e.g., see [9,15–17], different kinds of array grammars, both isometric and non-isometric ones, have been proposed, motivated by many application problems such as character recognition (also confer [3]), cluster analysis of patterns, and so on. Isometric contextual array grammars were introduced in [7]. Isometric contextual array grammars with matrix control and with regular control were investigated in [5]. Non-isometric contextual array grammars (with regulation) were considered in [10]. A basic variant of non-isometric contextual array grammars (without using selectors) – without control and with regular control – has been considered in [4].

In this paper we consider the extension of this basic variant now in addition using local selectors, i.e., along the whole cross-section between the array generated so far and the adjoined rectangle locally context conditions in the usual sense of selector and context must be fulfilled. For $d > 1$ we can show that non-isometric d-dimensional contextual array grammars with regular control and local selectors can simulate the computations of $(d-1)$-dimensional array grammars or Turing machines. Hence, for example, the emptiness problem for non-isometric d-dimensional contextual array grammars with regular control and local selectors for $d > 1$ is undecidable. On the other hand, in the 1-dimensional case we show that the string images of the array languages generated by non-isometric 1-dimensional contextual array grammars with regular control and local selectors or by isometric 1-dimensional contextual array grammars with regular control characterize the linear languages.

2 Definitions

For notions and notations as well as results related to formal language theory, we refer to textbooks like [18,19]; in particular, λ denotes the empty string. For the definitions and notations for arrays and sequential isometric array grammars we refer to [6,16,22]. The families of (λ-free) arbitrary, monotone, context-free, linear, and regular string languages over a k-letter alphabet are denoted by $\mathcal{L}(ENUM^k)$, $\mathcal{L}(MON^k)$, $\mathcal{L}(CF^k)$, $\mathcal{L}(LIN^k)$, and $\mathcal{L}(REG^k)$, respectively; we omit the superscript k if the size of the terminal alphabet may be arbitrary.

Let $\mathbb{Z}$ be the set of integers, let $\mathbb{N}$ be the set of positive integers. Let $d \in \mathbb{N}$. A d-*dimensional array* $\mathcal{A}$ over the alphabet V is a mapping $\mathcal{A} : \mathbb{Z}^d \rightarrow V \cup \{\#\}$ where $shape(\mathcal{A}) = \{v \in \mathbb{Z}^d \mid \mathcal{A}(v) \neq \#\}$ is finite and $\# \notin V$ is called the *blank symbol*. We usually write $\mathcal{A} = \{(v, \mathcal{A}(v)) \mid v \in shape(\mathcal{A})\}$. The set of all d-dimensional arrays over V is denoted by V^{*d}. The *empty array* Λ_d in V^{*d} satisfies $shape(\Lambda_d) = \emptyset$. Moreover, we define $V^{+d} = V^{*d} \setminus \{\Lambda_d\}$.

Let $v \in \mathbb{Z}^d$. Then the *translation* $\tau_v : \mathbb{Z}^d \rightarrow \mathbb{Z}^d$ is defined by $\tau_v(w) = w + v$ for all $w \in \mathbb{Z}^d$, and for any array $\mathcal{A} \in V^{*d}$ we define $\tau_v(\mathcal{A})$, the corresponding d-dimensional array translated by v, by $(\tau_v(\mathcal{A}))(w) = \mathcal{A}(w - v)$ for all $w \in \mathbb{Z}^d$. The vector $(0, ..., 0) \in \mathbb{Z}^d$ is denoted by Ω_d; the unit vector in $\mathbb{Z}^d$ with all components being 0 except the i-th one which is 1 is denoted by $e_{d,i}$.

Usually (see [16]) arrays are regarded as equivalence classes of arrays with respect to linear translations. The equivalence class $[\mathcal{A}]$ of an array $\mathcal{A} \in V^{*d}$ satisfies $[\mathcal{A}] = \{\mathcal{B} \in V^{*d} \mid \mathcal{B} = \tau_v(\mathcal{A}) \text{ for some } v \in \mathbb{Z}^d\}$. The set of all equivalence classes of d-dimensional arrays over V with respect to linear translations is denoted by $[V^{*d}]$, and this bracket notation carries over to classes of array languages as well.

As many results for d-dimensional arrays for a special d can be taken over immediately for higher dimensions, we introduce special notions:

Let $n, m \in \mathbb{N}$ with $n \leq m$. For $n < m$, the natural embedding $i_{n,m} : \mathbb{Z}^n \to \mathbb{Z}^m$ is defined by $i_{n,m}(v) = (v, \Omega_{m-n})$ for all $v \in \mathbb{Z}^n$; for $n = m$ we define $i_{n,n} : \mathbb{Z}^n \to \mathbb{Z}^n$ by $i_{n,n}(v) = v$ for all $v \in \mathbb{Z}^n$. To an n-dimensional array $\mathcal{A} \in V^{+n}$ with $\mathcal{A} = \{(v, \mathcal{A}(v)) \mid v \in shape(\mathcal{A})\}$ we assign the m-dimensional array $i_{n,m}(\mathcal{A}) = \{(i_{n,m}(v), \mathcal{A}(v)) \mid v \in shape(\mathcal{A})\}$.

We can use the well-known graph-theoretic notion of a connected graph to define connected arrays. Let W be a non-empty finite subset of $\mathbb{Z}^d$. We associate a graph $g(W)$ to W with vertex set W and an edge between $v, w \in W$ if and only if $\|v - w\| = 1$, where the norm $\|u\|$ of a vector $u \in \mathbb{Z}^d$, $u = (u(1), ..., u(d))$, is defined by $\|u\| = \max\{|u(i)| \mid 1 \leq i \leq d\}$. Then W is said to be *connected* if $g(W)$ is connected. There is a natural bijection between the equivalence classes of 1-dimensional connected arrays and strings: for any 1-dimensional array $\mathcal{A} = \{[((i-1), a_i)] \mid 1 \leq i \leq n\}$ we define its *string image* as $str(\mathcal{A}) = \{a_1 \ldots a_n\}$, and the string $w = a_1 \ldots a_n$ can be interpreted as the array $arr(w) = \{\{((i-1), a_i)\} \mid 1 \leq i \leq n\}$. In the standard way, these notions can be extended from strings and arrays to sets of strings and arrays.

In this paper, we will only consider a d-*dimensional array production* p over an alphabet V, $V = N \cup T$, where N is the alphabet of *non-terminal symbols*, T is the alphabet of *terminal symbols*, $N \cap T = \emptyset$, $\# \notin N \cup T$, to be in the 2-*normal form*: either $p = AvB \to CD$ with $A, B, C, D \in V \cup \{\#\}$, $A \neq \#$, $v \in \mathbb{Z}^d$ and $\|v\| = 1$ or $p = A \to b$ with $A \in N$ and $b \in T$. $AvB \to CD$ can be applied to a d-dimensional array $\mathcal{C}_1$ if and only if $\mathcal{C}_1$ contains a subarray $\{(u, A), (u+v, B)\}$, and the result of the application is an array $\mathcal{C}_2$ where this subarray has been replaced by $\{(u, C), (u+v, D)\}$. The application of $p = A \to b$ means replacing A by b. In both cases, we also write $\mathcal{C}_1 \Longrightarrow_p \mathcal{C}_2$. Moreover we say that the array $\mathcal{B}_2 \in [V^{*d}]$ is *directly derivable* from the array $\mathcal{B}_1 \in [V^{+d}]$ by the d-dimensional array production p if and only if there exist $\mathcal{C}_1 \in \mathcal{B}_1$ and $\mathcal{C}_2 \in \mathcal{B}_2$ such that $\mathcal{C}_1 \Longrightarrow_p \mathcal{C}_2$; we also write $\mathcal{B}_1 \Longrightarrow_p \mathcal{B}_2$.

Let us illustrate this concept with a small example:

$$\cdots \boxed{\#} \boxed{\#} \boxed{a} \boxed{X} \boxed{a} \boxed{b} \boxed{X} \boxed{\#} \boxed{\#} \cdots$$

denotes a connected 1-dimensional array $aXabX$, surrounded by blank symbols $\#$. The production $X(1)\# \to ba$ is applicable in this array, yet only to the rightmost variable X, and its application yields:

$$\cdots \boxed{\#} \boxed{\#} \boxed{a} \boxed{X} \boxed{a} \boxed{b} \boxed{b} \boxed{a} \boxed{\#} \cdots$$

Observe that the production $X(1)\# \to ba$ is not applicable to the leftmost variable X in this array, in contrast to the string case where the context-free production $X \to ba$ would also be applicable to the leftmost variable X in the string $aXabX$.

We call a d-dimensional array production $p = AvB \to CD$ in P

- *monotone*, if C and D are non-blank symbols from V;
- *strictly context-free*, if it is monotone as well as, moreover, $A \in N$ and $B = \#$;
- *regular*, if p is strictly context-free, and moreover, $C \in T$ and $D \in N$.

A terminal array production $A \to b$ is defined to be regular and therefore strictly context-free and monotone, too.

The array production $X(1)\# \to ba$ from our previous example is strictly context-free, but not regular.

Definition 1. *A d-dimensional array grammar is a quintuple*

$$G = (d, N, T, \#, P, S),$$

where N is the alphabet of non-terminal symbols, T is the alphabet of terminal symbols, $N \cap T = \emptyset$, $\# \notin N \cup T$; P is a finite non-empty set of d-dimensional array productions over $N \cup T$, and $S \in N$ is the start symbol.

We say that the array $\mathcal{C}_2 \in V^{*d}$ is *directly derivable* from the array $\mathcal{C}_1 \in V^{+d}$ in G, denoted $\mathcal{C}_1 \Longrightarrow_G \mathcal{C}_2$, if and only if there exists a d-dimensional array production $p \in P$ such that $\mathcal{C}_1 \Longrightarrow_p \mathcal{C}_2$. Let $\Longrightarrow_G^*$ be the reflexive transitive closure of $\Longrightarrow_G$. The *d-dimensional array language generated by G*, $L(G)$, is defined by

$$L(G) = \left\{ \mathcal{C} \in T^{+d} \mid \{(\Omega_d, S)\} \Longrightarrow_G^* \mathcal{C} \right\}.$$

For equivalence classes of arrays we say that the array $\mathcal{B}_2 \in \left[V^{*d}\right]$ is *directly derivable* from the array $\mathcal{B}_1 \in \left[V^{+d}\right]$ in G, denoted $\mathcal{B}_1 \Longrightarrow_G \mathcal{B}_2$, if and only if there exist $\mathcal{C}_1 \in \mathcal{B}_1$ and $\mathcal{C}_2 \in \mathcal{B}_2$ such that $\mathcal{C}_1 \Longrightarrow_p \mathcal{C}_2$. Then the d-*dimensional array language generated by G*, $[L(G)]$, is defined by

$$[L(G)] = \left\{ \mathcal{B} \in \left[T^{+d}\right] \mid [\{(\Omega_d, S)\}] \Longrightarrow_G^* \mathcal{B} \right\}.$$

G is called to be of type d-$ENUMA$, d-$MONA$, d-$SCFA$, and d-$REGA$ if all d-dimensional array productions in P are arbitrary, monotone, strictly context-free or regular, respectively. The corresponding families of Λ-free d-dimensional array languages of arrays and of equivalence classes of arrays over a k-letter alphabet are denoted by $\mathcal{L}\left(d\text{-}ENUMA^k\right)$, $\mathcal{L}\left(d\text{-}MONA^k\right)$, $\mathcal{L}\left(d\text{-}SCFA^k\right)$, and $\mathcal{L}\left(d\text{-}REGA^k\right)$ and by $\left[\mathcal{L}\left(d\text{-}ENUMA^k\right)\right]$, $\left[\mathcal{L}\left(d\text{-}MONA^k\right)\right]$, $\left[\mathcal{L}\left(d\text{-}SCFA^k\right)\right]$, and $\left[\mathcal{L}\left(d\text{-}REGA^k\right)\right]$, respectively; for arbitrary alphabets, we omit the superscript k.

In this paper we are only interested in (languages of) connected arrays, which for monotone array grammars is already guaranteed if we restrict ourselves to the 2-normal form. Thus, in the following we assume that in array languages in the families defined above only connected arrays are contained. These families of array languages form a Chomsky-like hierarchy; the following results are already folklore:

Theorem 1. *For all $d \geq 2$ and all $k \geq 1$,*

$$\left[\mathcal{L}\left(d\text{-}ENUMA^k\right)\right] \underset{\neq}{\supseteq} \begin{array}{l}\left[\mathcal{L}\left(d\text{-}MONA^k\right)\right] \\ \left[\mathcal{L}\left(d\text{-}SCFA^k\right)\right]\end{array} \underset{\neq}{\supseteq} \left[\mathcal{L}\left(d\text{-}REGA^k\right)\right].$$

Theorem 2. *For all $k \geq 1$,*

$$\left[\mathcal{L}\left(1\text{-}SCFA^k\right)\right] = \left[\mathcal{L}\left(1\text{-}REGA^k\right)\right] = \left[arr\left(\mathcal{L}\left(REG^k\right)\right)\right] \text{ and}$$
$$str\left(\left[\mathcal{L}\left(1\text{-}REGA^k\right)\right]\right) = str\left(\left[\mathcal{L}\left(1\text{-}SCFA^k\right)\right]\right) = \mathcal{L}\left(REG^k\right).$$

For the technical proof of Theorem 7, we need a special normal form for arbitrary d-dimensional array grammars, which we call *one-pebble normal form* as any intermediate sentential form contains exactly one marked (barred) nonterminal symbol and only array productions involving this marked symbol can be applied; this way of applying productions resembles the working mode of d-dimensional Turing machines with the barred symbol marking the position of the read/write head of the machine:

Lemma 1 *(One-pebble normal form).* *For any d-dimensional array grammar $G = (d, N, T, \#, P, S)$ in 2-normal form we can effectively construct an equivalent d-dimensional array grammar $G' = \left(d, N' \cup \bar{N}', T, \#, P', \bar{S}\right)$ such that every rule in P' is of the form $\bar{A}vB \to C\bar{D}$, with $\bar{A}, \bar{D} \in \bar{N}'$, $B, C \in N' \cup T \cup \{\#\}$, or $\bar{A} \to b$, with $\bar{A} \in \bar{N}'$, $b \in T \cup \{\#\}$.*

3   Isometric Contextual Array Grammars

In this section we repeat the main definitions and results on isometric contextual array grammars as developed in [5].

3.1   The Basic Variant

Definition 2. *A d-dimensional contextual array grammar $(d \in \mathbb{N})$ is a construct $G = (d, V, \#, P, A)$ where V is an alphabet not containing the blank symbol $\#$, A is a finite set of axioms, i.e., of d-dimensional arrays in V^{+d}, and P is a finite set of rules of the form $(U_\alpha, \alpha, U_\beta, \beta)$ where*

(i) $U_\alpha, U_\beta \subseteq \mathbb{Z}^d$, $U_\alpha \cap U_\beta = \emptyset$, and U_α, U_β are finite and non-empty;
(ii) $\alpha : U_\alpha \to V$ and $\beta : U_\beta \to V$.

(U_α, α) *corresponds with the selector and (U_β, β) with the context of the production $(U_\alpha, \alpha, U_\beta, \beta)$; U_α is called the selector area, and U_β the context area. As the sets U_α and U_β are uniquely determined by α and β, we will also represent $(U_\alpha, \alpha, U_\beta, \beta)$ by (α, β) only.*

For $\mathcal{C}_1, \mathcal{C}_2 \in V^{+d}$ we say that $\mathcal{C}_2$ is directly derivable from $\mathcal{C}_1$ by the contextual array production $p \in P$, $p = (U_\alpha, \alpha, U_\beta, \beta)$ (we write $\mathcal{C}_1 \Longrightarrow_p \mathcal{C}_2$), if there exists a vector $v \in \mathbb{Z}^d$ such that

- $C_1(w) = C_2(w) = \alpha(\tau_{-v}(w))$ for all $w \in \tau_v(U_\alpha)$,
- $C_1(w) = \#$ for all $w \in \tau_v(U_\beta)$,
- $C_2(w) = \beta(\tau_{-v}(w))$ for all $w \in \tau_v(U_\beta)$,
- $C_1(w) = C_2(w)$ for all $w \in \mathbb{Z}^d \setminus \tau_v(U_\alpha \cup U_\beta)$.

Hence, if in C_1 we find a sub-pattern that corresponds with the selector α and only blank symbols at the places corresponding with β, we can add the context β thus obtaining C_2.

$C_1 \Longrightarrow_G C_2$ means that $C_1 \Longrightarrow_p C_2$ for some $p \in P$. The array language generated by G is defined as

$$L(G) = \{C \in V^{+d} \mid A \Longrightarrow_G^* C \text{ for some } A \in \mathcal{A}\}.$$

The special type of d-dimensional contextual array grammars where axioms are connected and rule applications preserve connectedness is denoted by d-$ContA$, the corresponding family of d-dimensional array languages by $\mathcal{L}(d\text{-}ContA)$; by $\mathcal{L}(d\text{-}ContA^k)$ we denote the corresponding family of d-dimensional array languages over a k-letter alphabet.

3.2 Contextual Array Grammars with Regular Control

Following the general framework established in [8], for any type of grammars generating strings or arrays, we can define regular control as follows:

Definition 3. *Let G be a grammar of type X for generating strings or arrays with P being the set of productions in G. A pair $G_C = (G, L)$ with L being a regular string language over P is called a grammar with regular control. Derivations in G_C are defined as in G except that in a successful derivation the sequence of applied rules has to be a string from L. The language (of strings or arrays, respectively) generated by G_C is the set of all terminal objects which can be derived from (any of) the axiom(s) following a control string from L. The family of (string, array) languages generated by grammars of type X with regular control is denoted by $\mathcal{L}((X, REG))$.*

For example, the families of d-dimensional array languages of arrays and of equivalence classes of arrays genertted by d-dimensional contextual array grammars over a k-letter alphabet with regular control are denoted by $\mathcal{L}((d\text{-}ContA^k, REG))$ and $[\mathcal{L}((d\text{-}ContA^k, REG))]$, respectively.

The following results are among those established in [5]:

Theorem 3.

$$[\mathcal{L}(1\text{-}REGA^1)] = [\mathcal{L}((1\text{-}ContA^1, REG))] = [\mathcal{L}(1\text{-}ContA^1)].$$

Theorem 3 says that even with the regulating mechanisms of regular control languages, in the case of 1-dimensional contextual array grammars over a one-letter alphabet we cannot go beyond regularity, i.e., beyond $[\mathcal{L}(1\text{-}REGA^1)]$.

Theorem 4. *For any $d \geq 1$ and any $k \geq 2$, we have:*

$$[\mathcal{L}(d\text{-}ContA^k)] \subsetneqq [\mathcal{L}((d\text{-}ContA^k, REG))] .$$

Example 1. Consider the non-regular string language $L_{nr} = \{a^n ba^n \mid n \geq 1\}$. Due to Theorem 2, there cannot exist an array grammar of type $1\text{-}REGA^2$ (or even $1\text{-}SCFA^2$) G such that $[L(G)] = [arr(L_n)]$. Yet take the 1-dimensional contextual array grammar with regular control $G_C = (G_{nr}, L)$ with

$$G_{nr} = (1, \{a, b\}, \#, \{p_l, p_r\}, \{aba\})$$

where the contextual array productions can be depicted with the symbols of the selector being enclosed in boxes:

$$p_l = a \boxed{a} \text{ and } p_r = \boxed{a} a .$$

With the control language $L = \{p_l p_r\}^*$ we get $[L(G_C)] = [arr(L_{nr})]$. The control strings in L guarantee that the number of symbols a grows to the left and to the right in a synchronized way.

In sum, we get $[i_{1,d}(arr(L_n))] \in$

$$[\mathcal{L}((d\text{-}ContA^2, REG))] \setminus ([\mathcal{L}(d\text{-}ContA^2)] \cup [\mathcal{L}(d\text{-}SCFA^2)])$$

for any $d \geq 1$.

4   Non-isometric Contextual Array Grammars

A model of non-isometric contextual array grammars was considered in [10] and extended to arbitrary dimensions in [4]. In this paper, we will add local selectors to the rules to allow for controlling how the adjoined rectangles fit to the array already generated along the cross-section between this array and the adjoined rectangle.

We first define d-dimensional contextual array productions over a family of array languages $\mathcal{F}$.

Definition 4. *For any $d \geq 1$, the set of all d-dimensional rectangular arrays is denoted by $d\text{-}RECT$; the set of all d-dimensional rectangular arrays with length 1 in direction $e_{d,i}$ is denoted by $d\text{-}RECT_i$; any d-dimensional rectangular array in $d\text{-}RECT_i$ can also be considered as a $(d-1)$-dimensional rectangular array in $(d-1)\text{-}RECT$ by omitting component i.*

Definition 5. *A non-isometric d-dimensional contextual array production p over a family of $(d-1)$-dimensional array languages $\mathcal{F}$ is of the form $vL_1 \ldots L_m$ where $v \in \{e_{d,i}, -e_{d,i}\}$, $m \geq 1$, $1 \leq i \leq d$, and the L_j, $1 \leq j \leq m$, are taken from $\mathcal{F} \cap (d-1)\text{-}RECT$. For $d = 1$, the family $\mathcal{F}$ of $(d-1)$-dimensional array languages, i.e., of "0-dimensional" array languages, consists of all finite sets of symbols, and this family will also be denoted by $\mathcal{F}_0$. Moreover, to formally* complete the definitions, 0-RECT is understood to coincide with $\mathcal{F}_0$, which in fact means that it consists of the "0-dimensional" squares that are single symbols at the origin (0).

Then the application of a d-dimensional contextual array production $p = vL_1 \dots L_m$ to a d-dimensional array $\mathcal{A} \in$ d-RECT in the external mode is defined as yielding any d-dimensional array obtained by adjoining $\mathcal{A}_1 \dots \mathcal{A}_m$, where $\mathcal{A}_j \in L_j$, $1 \le j \le m$, on the border of $\mathcal{A}$ in the direction v such that the resulting array is again rectangular.

Remark 1. Notice that we have used a kind of catenation notation within the definition of $p = vL_1 \dots L_m$ to denote the adjoining of rectangles that fit together. Let us explain what this means for the important special case $d = 2$. Then, $L_1, \dots, L_m$ are 1-dimensional array languages (which can likewise be interpreted as string languages). What happens if p is applied to some array $\mathcal{C}$ of width ℓ and height h? If $v = e_{2,1}$, then we have to select strings $w_1, \dots, w_m$ of length h such that $w_i \in str(L_i)$. Together, these strings form an array $\mathcal{A}$ of width m and height h. More formally, w_i corresponds to a 2-dimensional array $\mathcal{A}_i$ of height h and width 1. Using the classical column catenation operation $①$, applying p to $\mathcal{A}$, with the chosen arrays $\mathcal{A}_1, \dots, \mathcal{A}_m$, results in

$$\mathcal{C} ① \mathcal{A}_1 ① \cdots ① \mathcal{A}_m.$$

If we take $v = -e_{2,1}$, staying with the choice of $\mathcal{A}_i$ as explained above, we obtain

$$\mathcal{A}_m ① \cdots ① \mathcal{A}_1 ① \mathcal{C}.$$

In a similar way, taking $v = e_{2,2}$ or $v = -e_{2,2}$ results in interpreting the adjoining operation (written like the usual catenation) as the classical row catenation $\ominus$.

Remark 2. In contrast to the definitions given in [4], we here only adjoin rectangles into one of the directions $e_{d,i}$ or $-e_{d,i}$ and not at both "sides" at the same time, as we will use regular control afterwards. In particular, this means that we can simulate the contextual grammars as defined in [4] by using regular control within the framework introduced in this paper.

Definition 6. *A (non-isometric) d-dimensional contextual array grammar is a construct $G = (d, V, \#, P, A)$ where V is a finite alphabet of symbols, $\# \notin V$ is the blank symbol, P is a finite non-empty set of non-isometric d-dimensional contextual array productions over $\mathcal{F}$, and A is the finite set of axioms (d-dimensional arrays). The families of d-dimensional array languages generated by non-isometric d-dimensional contextual array grammars in the external mode are denoted by $[\mathcal{L}(d\text{-}ExtContA(\mathcal{F}))]$.*

Example 2. Consider the regular string language $L_{line} = \{a^n \mid n \ge 1\}$. Due to Theorem 2, $[arr(L_{line})] \in [\mathcal{L}(1\text{-}REGA^1)]$, but we also have

$$[arr(L_{line})] \in [\mathcal{L}(1\text{-}ExtContA(\mathcal{F}_0))]$$

as $[arr\,(L_{line})]$ can be generated by the 1-dimensional non-isometric contextual array grammar G_{line} with

$$G_{line} = (1, \{a\}, \#, \{p_{line}\}, \{arr\,(a)\})$$

and $p_{line} = (1)\,\{a\}$. The singleton set $\{a\}$ can be represented by a only, hence, we can also write $p_{line} = (1)a$. $\qquad\qquad\qquad\qquad\qquad\qquad\qquad$ □

4.1 Non-isometric Contextual Array Grammars with Regular Control

We now again add regular control to the model of non-isometric contextual array grammars:

Definition 7. *A* non-isometric *d*-dimensional contextual array grammar with regular control *is a pair* $G_C = (G, L)$ *where* $G = (d, V, \#, P, A)$ *is a non-isometric d-dimensional contextual array grammar and* L *is a regular string language over* P, *i.e.,* $L \subseteq P^*$.

Example 3. Consider $G = (2, \{a\}, \#, P, \{i_{i,2}\,(arr\,(a))\})$ with $P = \{p_u, p_r\}$, $p_u = (0, 1)\,L$, $p_r = (1, 0)\,L$, and $L = [arr\,(\{a^n \mid n \ge 1\})]$. Then $G_C = (G, \{p_u p_r\}^*)$ generates the set of all sqares over $\{a\}$.

Example 4. Let $L_x^d = [i_{1,d}\,(arr\,(L_x))]$ where $L_x = \{x^n \mid n \ge 1\}$ for $x \in \{a, b\}$, and consider $L_{a+b} = L_a \cup L_b$ which, obviously, is a regular string language and, interpreted as a language of 1-dimensional arrays, is regular, too, i.e., with defining $L_{a+b}^d = [i_{1,d}\,(arr\,(L_{a+b}))]$ we have $L_{a+b}^d \in [\mathcal{L}\,((d\text{-}REGA))]$. On the other hand, without having the chance to check in which context new columns should be inserted, even regular control languages cannot enable external contextual array grammars to generate L'_{a+b} as adding new symbols a can also be performed for underlying arrays so far only containing symbols b and vice versa, i.e.,

$$L_{a+b}^d \notin [\mathcal{L}\,((d\text{-}ExtContA\,(((d-1)\text{-}RECT)), REG))].$$

4.2 Non-isometric Contextual Array Grammars with Regular Control and Local Selectors

The drawback of the basic model of non-isometric contextual array grammars, even with regular control, as exhibited in Example 4, now leads us to extend the model with local selectors Sel to be assigned to the non-isometric contextual array productions $vL_1 \ldots L_m$ with $v \in \{e_{d,i}, -e_{d,i}\}$.

Definition 8. *A* non-isometric *d*-dimensional contextual array production *p* with local selectors *over a family of* $(d-1)$-dimensional array languages $\mathcal{F}$ *is of the form* (p', Sel) *where* $p' = vL_1 \ldots L_m$ *is a non-isometric d-dimensional contextual array production with* $v \in \{e_{d,i}, -e_{d,i}\}$, $m \ge 1$, $1 \le i \le d$, *and the* L_j, $1 \le j \le m$, *are taken from* $\mathcal{F} \cap (d-1)\text{-}RECT$ *and* Sel *is a finite set of local* selectors. *A local selector is a pair (s, c) where both s and c are d-dimensional arrays which put together in direction v fit to form a d-dimensional rectangle.*

The application of a d-dimensional contextual array production with local selectors to a d-dimensional array $\mathcal{A} \in d\text{-}RECT$ in the external mode is defined as for applying p', yet in addition the constraints given by the set of selectors Sel have to be fulfilled, i.e., along the whole cross-section between the array derived so far and the newly adjoined rectangle one of the selectors from Sel must fit, i.e., putting together (s, c) in direction v, s is a subarray of $\mathcal{A}$ and c is a subarray of the adjoined rectangle.

Remark 3. As explained in Remark 1, the formal treatment of what it means to put together d-dimensional rectangular arrays so that they form a rectangle is rather intricate. However, the intuition behind should become clear with giving an example for the 2-dimensional case: For an alphabet V, the condition to only adjoin the same symbols along the whole line, we can use the set of selectors $\{(a, a) \mid a \in V\}$.

Definition 9. *A non-isometric d-dimensional contextual array grammar with regular control and local selectors is a pair $G_C = (G, L)$ where $G = (d, V, \#, P, A)$ is a non-isometric d-dimensional contextual array grammar with d-dimensional contextual array production with local selectors in P and L is a regular string language over P, i.e., $L \subseteq P^*$. The family of d-dimensional array languages of equivalence classes of arrays generated by these array grammars over a k-letter alphabet is denoted by $\left[\mathcal{L}\left(\left(d\text{-}ExtContASel^k\left((d-1)\text{-}RECT \right), REG \right) \right) \right]$. If $L = P^*$, we call G a non-isometric d-dimensional contextual array grammar with local selectors, the corresponding family of languages of rectangular arrays is denoted by $\left[\mathcal{L}\left(d\text{-}ExtContASel^k\left((d-1)\text{-}RECT \right) \right) \right]$.*

We now return back to Example 4:

Example 5. $\left[arr\left(L_{a+b} \right) \right]$ now can be generated by $G = (1, \{a, b\}, \#, P, \{a, b\})$ where P contains the rules $((1)a, \{(a, a)\})$ and $((1)b, \{(b, b)\})$, i.e., adjoining $x \in \{a, b\}$ with the rule $(1)x$ is controlled by the selector (x, x). Hence, we get $\left[arr\left(L_{a+b} \right) \right] \in \left[\mathcal{L}\left(1\text{-}ExtContASel^2\left((d-1)\text{-}RECT \right) \right) \right]$.

In the 1-dimensional case, isometric d-dimensional contextual array grammars with regular control and non-isometric d-dimensional contextual array grammars with regular control and local selectors in fact are not distinguishable, i.e., we have the rather astonishing result:

Theorem 5. *For any $k \geq 1$,*

$$\left[\mathcal{L}\left(\left(1\text{-}ExtContASel^k\left(\mathcal{F}_0 \right), REG \right) \right) \right] = \left[\mathcal{L}\left(\left(1\text{-}ContA^k, REG \right) \right) \right].$$

Proof. (Sketch) A contextual array production q in a 1-dimensional contextual array grammar inserting $b_1 \ldots b_n$ in the context of $a_1 \ldots a_m$ can be depicted as $\boxed{a_1} \cdots \boxed{a_m} b_1 \ldots b_n$ for rules growing the underlying 1-dimensional array (string) to the right, i.e., into direction (1)) and $b_n \ldots b_1 \boxed{a_1} \cdots \boxed{a_m}$ for rules growing the underlying 1-dimensional array (string) to the left (into direction (-1)). According to the definition of $\mathcal{F}_0$, its elements in fact can simply be interpreted as symbols over V. A 1-dimensional contextual array production with local selectors therefore is of the form (p, Sel) with $p = vb_1 \ldots b_n$, $v \in \{(1), (-1)\}$, and the elements Sel being of the form $s = (a_1 \ldots a_m, b_1 \ldots b_n)$ for $v = (1)$ and $s = (a_1 \ldots a_m, (-1) b_1 \ldots (-n) b_n)$ for $v = (-1)$. Hence, the effect of applying q is the same as the effect of the application of $(p, \{s\})$. Finally we observe that the selectors in Sel can be handled independently from each other, as in the special situation of the 1-dimensional case for checking the context only one selector can be used.   □

5   A Characterization of Linear String Languages

In this section we will derive a characterization of $\mathcal{L}(LIN)$ in terms of 1-dimensional array grammars and (isometric and non-isometric) contextual array grammars. It is well-known that the emptiness problem for the corresponding class of linear grammars can be solved effectively, so that the constructiveness of our proofs yields a corresponding decidability result for the many types of array grammars that we have discussed in this paper.

More precisely, we are going to prove that $\mathcal{L}(LIN) =$

$$str([\mathcal{L}((1\text{-}ContA, REG))]) = str([\mathcal{L}((1\text{-}ExtContASel^k(\mathcal{F}_0), REG))]).$$

Due to Theorem 5, we only have to show

$$str([\mathcal{L}((1\text{-}ContA, REG))]) = \mathcal{L}(LIN).$$

The backbone of the stated characterization is the following normal form result:

Lemma 2. *For any 1-dimensional contextual array grammar $G_C = (G, L)$ with regular control, where $G = (1, V, \#, P, A)$, $L \subseteq P^*$, we can construct an equivalent 1-dimensional contextual array grammar $G'_C = (G', L')$ with regular control, $G = (1, V, \#, P', A')$, $L' \subseteq P'^*$, such that the rules in P' have the following two properties:*

- *All rules are either of the form $\boxed{a}\,b$ or of the form $b\,\boxed{a}$ for some $a, b \in V$.*
- *If there is a rule of the form $\boxed{a}\,b$ (or $b\,\boxed{a}$, respectively) in P', then also all rules of the form $\boxed{c}\,b$ or $b\,\boxed{c}$, respectively, are in P', for any $c \in V$.*

Proof. The proof of this normal form result proceeds in several stages that we only sketch in the following.

1. As $\mathcal{L}(REG)$ is closed under union, it is easy to see that for any two contextual array grammars with regular control $G_C^{(i)} = (G, L^{(i)})$, $1 \leq i \leq 2$, and $G_C = (G, L^{(1)} \cup L^{(2)})$ we have $[L(G_C)] = \left[L\left(G_C^{(1)}\right)\right] \cup \left[L\left(G_C^{(2)}\right)\right]$. Moreover, by definition, for any contextual array grammar with regular control $G = (1, V, \#, P, A)$, $L \subseteq P^*$, we have $[L(G_C)] = \bigcup_{A \in A} [L(G_A)]$ where $G_A = (1, V, \#, P, \{A\})$.

2. If there is any rule whose context contains more than one letter, then it is possible to "spell out" this context; recall that this reduces to spelling out a string, and this linear task can easily be controlled by slightly modifying the control language; the resulting control language again is regular. We leave the details of this construction to the reader, but this is standard from many similar constructions in basic formal language theory. For simplicity, in the following we refer to the grammar obtained in this way again as $G_C = (G, L)$ with $G = (1, V, \#, P, A)$, $L \subseteq P^*$. Moreover, without loss of generality, we may assume that A contains only one axiom, i.e., $A = \{\mathcal{A}\}$, as $\mathcal{L}(LIN)$ is closed under union, too.

3. We have to reduce the size of the selector; in fact, as formalized in the second condition of the normal form, the grammar will only be able to see if the task is to append something to the left or to the right of the current 1-dimensional array (or string, in the finally intended interpretation).

   Let s be the length of the longest selector of any rule in P. In L_s, we collect all 1-dimensional arrays of length at most $\max\{2s+2, |str(\mathcal{A})|\}$ derivable from $\mathcal{A}$. In a first phase, we spell out all (finitely many) strings in $V^{\leq s}$ and keep track of the state that would have been reached by inputting the corresponding control string into the finite automaton accepting L while producing each string by G_C (if it is producible at all).

   Due to the properties ensured so far, we can store the leftmost and the rightmost s letters of the 1-dimensional array obtained so far in the state information of the finite automaton that accepts L'. More formally, the state set Q' of the automaton accepting L' is a superset of $Q \times V^s \times V^s$, where Q is the state set of a finite automaton accepting L. Therefore, whenever we are going to simulate a rule application of G_C that first checks the context of s symbols to the left or to the right, we find this information encoded in the state set Q', and the simulation therefore can proceed correctly.

Hence, we can guarantee the promised properties, as possibly adding some rules to P' that are never used in control strings does no harm.   □

Having derived this normal form, we can proceed with the proof of the desired characterization theorem.

Theorem 6. $\mathcal{L}(LIN) = str([\mathcal{L}((1\text{-}ContA, REG))])$
$$= str([\mathcal{L}((1\text{-}SCFA, REG))]).$$

Proof. For showing $str([\mathcal{L}((1\text{-}ContA, REG))]) \subseteq \mathcal{L}(LIN)$ it suffices to point to the proof of Theorem 12.9 in [14], because especially the second property of the normal form lemma means that – translated to the string case – we are not using any choice at all, i.e., each language in $str([\mathcal{L}((1\text{-}ContA, REG))])$ is a KEC language, to make use of the terminology of Gheorghe Păun. Hence, it is also a linear language.

To show the converse, we can proceed similar to the proof of Theorem 12.9 in [14]. This means that the nonterminal information of the linear grammar is stored in the state of the finite control of the control language. However, we have to split the simulation of a linear rule $A \to uBv$ into two steps. First, we adjoin u to the left, and then we adjoin v to the right. The correct sequence of these to adjoining steps again can easily be controlled by the regular control language.

For showing $str([\mathcal{L}((1\text{-}SCFA, REG))]) = \mathcal{L}(LIN)$ we observe that arrays (strings) being derived by a 1-dimensional strictly context-free array grammar with regular control grow to the left and to the right only guided by the control language, hence, similar arguments as above apply. This observation concludes the proof of our characterization theorem. $\square$

We remark that the construction somehow resembles the way how 2-head finite automata can accept the linear languages: the regular control is essentially the finite state control and the two extremes in which a string can be extended in the present model can be identified by the two heads that read the input from the two extremes. This technique has been employed in some arguments in [12]. Similar constructions can be also found in [2,20,21].

Corollary 1. *The emptiness problem is decidable for 1-dimensional contextual array grammars with regular control.*

Proof. As the emptiness problem is decidable for linear string languages, the claim immediately follows from Theorem 6, because all proof steps explained there are constructive. $\square$

As we will see in the following section, this decidability result only holds for the 1-dimensional case, whereas for $d \geq 2$ in general the emptiness problem is undecidable for d-dimensional non-isometric contextual array grammars with regular control and local selectors.

6   More-Dimensional Non-isometric Contextual Array Grammars with Regular Control and Local Selectors

In this section we will exhibit that, for $d \geq 2$, d-dimensional non-isometric contextual array grammars with regular control and local selectors can simulate the computations of $(d-1)$-dimensional array grammars; as a consequence of this result we infer the undecidability of the emptiness problem for d-dimensional non-isometric contextual array grammars with regular control and local selectors.

Theorem 7. *For any $d \geq 2$, the computations of any $(d-1)$-dimensional array grammar G can be simulated by a d-dimensional non-isometric contextual array grammars with regular control and local selectors $G_C = (G', L)$ in such a way that in each derivation step of G_C a $(d-1)$-dimensional rectangle representing the next sentential form obtained in the derivation of G is adjoined.*

Proof. The main idea of the proof is to adjoin another $(d-1)$-dimensional "slice", i.e., $(d-1)$-dimensional rectangle to the current d-dimensional array which represents the derivation carpet generated so far of the $(d-1)$-dimensional array grammar to be simulated. The construction then proceeds in three stages:

1. First, a $(d-1)$-dimensional rectangle of symbols $X_\#$ representing the blank symbol $\#$ is generated, having somewhere in the middle the start symbol S of G and on the borders having special symbols as delimiters. This stage is guided by the initial part of the control language L that we denote by L_I.
2. In the main part, the array productions in the set of array productions P from G of the form $\bar{A}vB \to C\bar{D}$ are simulated by using specific selectors guaranteeing that only a new $(d-1)$-dimensional "slice" is adjoined in direction $e_{d,d}$ which represents the next sentential form of the derivation in G to be simulated; an important feature to keep the changes local is assuming G to be in the one-pebble normal form. The corresponding set of non-isometric contextual array productions with local selectors in G' is denoted by P_S.
3. Finally, the derivation in G ends with the application of a rule $\bar{A} \to b$; the set of non-isometric contextual array productions with local selectors in G' for simulating those is denoted by P_F.

In total, we obtain $L = L_I P_S^* P_F$.

We now explain the construction for $d = 2$ in more detail, and then explain how this construction generalizes to the d-dimensional case for $d > 2$.

Now let $G = \left(1, N \cup \bar{N}, T, \#, P, S\right)$ be a 1-dimensional array grammar in one-pebble normal from as defined in Lemma 1. The 2-dimensional non-isometric contextual array grammars with regular control and local selectors $G_C = (G', L)$ with $G' = \left(1, N \cup \bar{N}, T, \#, P', \bar{S}\right)$, is constructed as follows:

First, from the axiom $\left\{(0,0)\,D, (0,1)\,\bar{S}\right\}$ the rules with the local selectors

$$
s_{I,S}\quad
\begin{array}{cc}
X_\# & \boxed{\bar{S}} \\
D & \boxed{D}
\end{array},
\qquad
s_{I,S'} =
\begin{array}{cc}
\boxed{\bar{S}} & X_\# \\
\boxed{D} & D
\end{array},
$$

$$
s_{I,\#} =
\begin{array}{cc}
X_\# & \boxed{X_\#} \\
D & \boxed{D}
\end{array},
\qquad
s_{I,\#'} =
\begin{array}{cc}
\boxed{X_\#} & X_\# \\
\boxed{D} & D
\end{array},
$$

$$
s_{I,L} =
\begin{array}{cc}
L & \boxed{X_\#} \\
D & \boxed{D}
\end{array},
\qquad
s_{I,R} =
\begin{array}{cc}
\boxed{X_\#} & R \\
\boxed{D} & D
\end{array}.
$$

generate the double-line

$$
\begin{array}{l}
L\ X_\#\ \ldots\ X_\#\ \bar{S}\ X_\#\ \ldots\ X_\#\ R \\
D\ D\ \ldots\ D\quad D\ D\ \ldots\ D\quad D
\end{array}
$$

where the second line corresponds to the string $LX_\#^m \bar{S} X_\#^n R$, $m, n \geq 1$. The symbol $X_\#$ represents the blank symbol $\#$. With m and n being big enough, sufficient work space has to be generated. In the following, a blank symbol $\#$ in an array production from P has to be transformed into the corresponding symbol $X_\#$ in the rules of G' described below; yet in order to avoid too complicated notations, we will not distinguish between these two symbols below.

A rule $r \in P'$ of the 2-dimensional non-isometric contextual array grammars with regular control and local selectors G_C can be described by $r = (v_r, S_r)$ where $v \in \{(1,0), (-1,0), (0,1), (0,-1)\}$ is a vector indicating that a fitting vertical line (for $v \in \{(1,0), (-1,0)\}$) or horizontal line (for $v \in \{(0,1), (0,-1)\}$) over the alphabet $V' = N \cup \bar{N} \cup T \cup \{X_\#, D, L, R\}$ has to be adjoined obeying to the restrictions given by the local selectors in S_r.

For the initial rules described above, we define the initial set of rules

$$P_I = \{((-1,0), S_r) \mid S_r \in \{s_{I,S}, s_{I,\#}, s_{I,L}\}\}$$
$$\cup \{((1,0), S_r) \mid S_r \in \{s_{I,S'}, s_{I,\#'}, s_{I,R}\}\}$$

and the initial part of the control language

$$L_I = \{s_{I,S}\} \{s_{I,S'}\} \{s_{I,\#}, s_{I,\#'}\}^* \{s_{I,L}\} \{s_{I,R}\}.$$

A 1-dimensional array production $p = \bar{A}(1) B \to C\bar{D}$ from P is captured by the local selectors

$$s_{p,1} = \begin{array}{cc} E & C \\ \boxed{E} & \boxed{\bar{A}} \end{array}, \quad s_{p,2} = \begin{array}{cc} C & \bar{D} \\ \boxed{\bar{A}} & \boxed{B} \end{array}, \quad s_{p,3} = \begin{array}{cc} \bar{D} & E \\ \boxed{B} & \boxed{E} \end{array},$$

for any $E \in N \cup T \cup \{X_\#\}$, and a 1-dimensional array production $p = \bar{A}(-1) B \to C\bar{D}$ from P is captured by the local selectors

$$s_{p,1} = \begin{array}{cc} E & \bar{D} \\ \boxed{E} & \boxed{B} \end{array}, \quad s_{p,2} = \begin{array}{cc} \bar{D} & C \\ \boxed{B} & \boxed{\bar{A}} \end{array}, \quad s_{p,3} = \begin{array}{cc} C & E \\ \boxed{\bar{A}} & \boxed{E} \end{array};$$

obviously, for the 1-dimensional array production $p_E = \bar{A} \to b$ from P we simply can take the local selectors

$$s_{p_E,1} = \begin{array}{cc} E & b \\ \boxed{E} & \boxed{\bar{A}} \end{array} \quad \text{and} \quad s_{p_E,2} = \begin{array}{cc} b & E \\ \boxed{\bar{A}} & \boxed{E} \end{array}.$$

Due to the one-pebble normal form of G it is guaranteed that for the correctness of the simulation of applying the 1-dimensional array productions $\bar{A}(1) B \to C\bar{D}$ and $\bar{A}(-1) B \to C\bar{D}$ from P only the local selectors

$$\begin{array}{cc} C & \bar{D} \\ \boxed{\bar{A}} & \boxed{B} \end{array} \quad \text{and} \quad \begin{array}{cc} \bar{D} & C \\ \boxed{B} & \boxed{\bar{A}} \end{array}$$

are relevant.

In addition, for covering the rest of the line to be adjoined correctly, we have to take the local selectors

$$s_L = \begin{array}{cc} L & E \\ \boxed{L} & \boxed{E} \end{array} \quad \text{and} \quad s_R = \begin{array}{cc} E & R \\ \boxed{E} & \boxed{R} \end{array} \quad \text{as well as} \quad s_{E,F} = \begin{array}{cc} E & F \\ \boxed{E} & \boxed{F} \end{array}$$

for any $E, F \in N \cup T \cup \{X_\#\}$; only in the local environment of the barred nonterminal other selectors have to be applied than these ones checking the identity of the adjoined symbols with the ones in the preceding line, and we collect these selectors in the set S_g.

For any array production $p \in P$, we define the corresponding rule in P' as $((0,1), S_p)$ where the set of selectors S_p for p being of the form $\bar{A}(1) B \to C\bar{D}$ or $\bar{A}(-1) B \to C\bar{D}$ is

$$S_p = S_g \cup \{s_{p,1}, s_{p,2}, s_{p,3}\};$$

these rules altogether form the rule set P_S. For any final array production $p_E = \bar{A} \to b$ we take $S_{p_E} = S_g \cup \{s_{p_E,1}, s_{p_E,2}\}$, which gives the set of rules P_F. The simulation part of the control language therefore is $L_S = P_S^*$. The simulation of a derivation in G has to end with a rule in P_F. In total, for the control language L we obtain $L = L_I P_S^* P_F$.

For higher dimensions, a similar construction applies with an initial phase represented by a suitable regular control language L_I, an intermediate simulation phase represented as P_S^* and the final phase P_F so that again the control language will be $L = L_I P_S^* P_F$. The selectors to be used in this case are again squares of side length 2, but now of course of dimension d. $\square$

The overall construction described above resembles the simulation of the work of a Turing machine. However, doing this formally is far more tedious compared to our construction given within the framework of array grammars.

Corollary 2. *For $d \geq 2$, the emptiness problem is undecidable for d-dimensional contextual array grammars with regular control and local selectors.*

Proof. As the emptiness problem is undecidable for d-dimensional array grammars as is the emptiness problem for arbitrary string grammars, the claim follows immediately from Theorem 7, because all proof steps explained there are constructive. $\square$

Remark 4. Both features – regular control and local selectors – are needed to obtain this undecidability result. Without selectors, the contextual array productions are always applicable as long as we take the arrays to be adjoined from $(d-1)$-$RECT$, hence, in that case the generated array language is empty if and only if the control language is empty. On the other hand, local selectors alone without regular control are pure grammars, i.e., the generated language at least contains all axioms.

7   Conclusions

Extending the basic model of non-isometric d-dimensional contextual array grammars with using local selectors, we have obtained a powerful tool to describe the evolution of more-dimensional objects, as for example, 2-dimensional pictures or even 3-dimensional objects.

For $d > 1$, we have shown that the emptiness problem is undecidable for d-dimensional contextual array grammars with regular control and local selectors, based on the result that d-dimensional non-isometric contextual array grammars with regular control and local selectors can simulate the computations of $(d - 1)$-dimensional array grammars. In contrast to that, in the 1-dimensional case the string images of array languages generated by isometric contextual array grammars with regular control or generated by non-isometric contextual array grammars with regular control (and with or without local selectors) characterize the linear languages, which implies that the corresponding emptiness problem is decidable.

References

1. Ehrenfeucht, A., Păun, Gh., Rozenberg, G.: Contextual grammars and formal languages. In: Rozenberg, G., Salomaa, A. (eds.) Handbook of Formal Languages, vol. 2, pp. 237–293. Springer, Heidelberg (1997)
2. Fernau, H.: Even linear simple matrix languages: formal language properties and grammatical inference. Theor. Comput. Sci. **289**, 425–489 (2002)
3. Fernau, H., Freund, R.: Bounded parallelism in array grammars used for character recognition. In: Perner, P., Rosenfeld, A., Wang, P. (eds.) SSPR 1996. LNCS, vol. 1121, pp. 40–49. Springer, Heidelberg (1996)
4. Fernau, H., Freund, R., Siromoney, R., Subramanian, K.G.: Regulated contextual array grammars, to appear in Annals of the University of Bucharest (Mathematical Series)
5. Fernau, H., Freund, R., Siromoney, R., Subramanian, K.G.: Contextual array grammars with matrix and regular control (in preparation)
6. Freund, R.: Control mechanisms on #-context-free array grammars. In: Păun, Gh. (ed.) Mathematical Aspects of Natural and Formal Languages, pp. 97–137. World Scientific Publication, Singapore (1994)
7. Freund, R., Păun, Gh., Rozenberg, G.: Chapter 8: contextual array grammars. In: Subramanian, K.G., Rangarajan, K., Mukund, M. (eds.) Formal Models Languages and Applications. Series in Machine Perception and Articial Intelligence, vol. 66, pp. 112–136. World Scientific, Singapore (2007)
8. Freund, R., Kogler, M., Oswald, M.: A general framework for regulated rewriting based on the applicability of rules. In: Kelemen, J., Kelemenová, A. (eds.) Computation, Cooperation, and Life. LNCS, vol. 6610, pp. 35–53. Springer, Heidelberg (2011)
9. Giammarresi, D., Restivo, A.: Two-dimensional languages. In: Rozenberg, G., Salomaa, A. (eds.) Handbook of Formal Languages, pp. 215–267. Springer, Heidelberg (1997)
10. Krithivasan, K., Balan, M.S., Rama, R.: Array contextual grammars. In: Martín-Vide, C., Păun, Gh. (eds.) Recent Topics in Mathematical and Computational Linguistics, pp. 154–168. Editura Academiei Române, Bucureşti (2000)
11. Marcus, S.: Contextual grammars. Revue Roumaine de Mathématiques Pures et Appliquées **14**, 1525–1534 (1969)
12. Nagy, B.: On a hierarchy of $5' \rightarrow 3'$ sensing Watson-Crick finite automata languages. J. Logic Comput. **23**, 855–872 (2013)
13. Păun, Gh., Nguyen, X.M.: On the inner contextual grammars. Revue Roumaine de Mathématiques Pures et Appliquée **25**, 641–651 (1980)
14. Păun, Gh.: Marcus Contextual Grammars. Studies in Linguistics and Philosophy. Kluwer Academic Publishers, Dordrecht (1997)

15. Pradella, M., Cherubini, A., Crespi-Reghizzi, S.: A unifying approach to picture grammars. Inf. Comput. **209**, 1246–1267 (2011)
16. Rosenfeld, A.: Picture Languages. Academic Press, Reading (1979)
17. Rosenfeld, A., Siromoney, R.: Picture languages - a survey. Lang. Des. **1**, 229–245 (1993)
18. Rozenberg, G., Salomaa, A. (eds.): Handbook of Formal Languages, vol. 1-3. Springer, Heidelberg (1997)
19. Salomaa, A.: Formal Languages. Academic Press, Reading (1973)
20. Subramanian, K.G.: A note on regular controlled apical growth filamentous systems. Int. J. Comput. Inf. Sci. **14**, 235–242 (1985)
21. Takada, Y.: A hierarchy of language families learnable by regular language learning. Inf. Comput. **123**, 138–145 (1995)
22. Wang, P.S.-P.: Some new results on isotonic array grammars. Inf. Process. Lett. **10**, 129–131 (1980)

Universality of Graph-controlled Leftist Insertion-deletion Systems with Two States

Sergiu Ivanov[1] and Sergey Verlan[1,2](✉)

[1] Laboratoire d'Algorithmique, Complexité et Logique,
Université Paris Est – Créteil Val de Marne,
61, av. gén. de Gaulle, 94010 Créteil, France
sergiu.ivanov@u-pec.fr
[2] Institute of Mathematics and Computer Science,
Academy of Sciences of Moldova,
Academiei 5, Chisinau MD-2028, Moldova
verlan@u-pec.fr

Abstract. In this article, we consider leftist insertion-deletion systems, in which all rules have contexts on the same side, and may only insert or delete one symbol at a time. We start by introducing extended rules, in which the contexts may be specified as regular expressions, instead of fixed words. We then prove that leftist systems with such extended rules and two-state graph control can simulate any arbitrary 2-tag system. Finally, we show how our construction can be simulated in its turn by graph-controlled leftist insertion-deletion systems with conventional rules of sizes $(1,1,0;1,2,0)$ and $(1,2,0;1,1,0)$ (where the first three numbers represent the maximal size of the inserted string and the maximal size of the left and right contexts respectively, while the last three numbers provide the same information about deletion rules), which implies that the latter systems are universal.

1 Introduction

Abstract insertion and deletion operations are simple, yet powerful, special cases of string rewriting rules. Intuitively, insertion is adding a substring at a site having a specified left, right, or both contexts, while deletion is removing a substring from a site having a specified left, right, or both contexts. The precursor of insertion was context adjoining, first introduced by S. Marcus in the seminal paper [19] with a linguistic motivation, and then further developed in [24,25]. The modern definition of insertion was introduced in [6] in the form of semi-contextual grammars.

The works [7,8] defined insertion differently, by generalising Kleene's operations of concatenation and closure [15]. Indeed, insertion can be seen as concatenation which is allowed to happen anywhere in the string. Following a similar approach, the work [12] introduced the dual operation of deletion as a generalised quotient operation which does not necessarily happen at the ends of the string. The paper [14] first considers systems containing finite sets of insertion

© Springer International Publishing Switzerland 2015
J. Durand-Lose and B. Nagy (Eds.): MCU 2015, LNCS 9288, pp. 79–93, 2015.
DOI: 10.1007/978-3-319-23111-2_6

and deletion rules working together: insertion-deletion systems. Such a system works in generative mode: it sequentially applies insertion or deletion rules to one of its finitely many axioms; the generated language includes all the terminal words obtained in the process.

Exciting sources of motivation for studying insertion and deletion operations were found in biology [4,13,26,28]. A well known one is the theoretically conceived process of mismatched annealing of DNA, which effectively results in insertions or deletions of certain segments of the strands [26]. In the formal framework of insertion and deletion operations, such modifications to DNA strands are modelled by context-free rules, i.e. insertion and deletion rules which can be applied anywhere in the string. The expressive power of such context-free operations was studied in the article [20], for example, which shows that context-free insertion-deletion operations of sizes $(3,0,0;2,0,0)$ and $(2,0,0;3,0,0)$ can simulate arbitrary string rewriting rules and are thus computationally complete. For a detailed overview of the computational power of context-free insertion and deletion operations, the reader is referred to [29]. For surveys of results on insertion-deletion systems in general, we refer to [16,30].

Another biological phenomenon which can be seen as a sequence of insertions and deletions is RNA editing, which was discovered in some species of protozoa [1,2]. RNA editing consists in inserting or deleting fragments of messenger RNA, and is guided by an anchor segment always located on one side of the edited locus. This directly motivates the study of one-sided insertion-deletion systems, i.e. systems in which all rules must have the context on one and the same side. The works [17,18,21] investigate the power of such systems and give several computational completeness results as a function of the size of the rules, as well as describe some families of one-sided systems which are not computationally complete. For these families, additional control mechanisms can be considered which often increase the expressive power, for example, matrix control [23], or semi-conditional and random context control [10].

One of the most frequently discussed variant of controlled insertion and deletion are *graph-controlled* insertion-deletion systems (sometimes also called insertion-deletion P systems). The work [18] shows that five-state graph control increases the power of small one-sided insertion-deletion rules to computational completeness. In [5], this result is improved upon and four-state graph control is shown to suffice for generating all recursively enumerable languages. The article [11] considers insertion-deletion systems of sizes $(1,2,0;1,1,0)$ and $(1,1,0;1,2,0)$, and proves that adding three-state graph-control to them results in computationally complete devices.

In this paper, we focus on a special variant of one-sided insertion-deletion systems, introduced in [9]: *leftist* systems. In such systems, the rules can only insert or delete one symbol at a time, and must use contexts on the same side. In [9], it is shown that all leftist insertion-deletion systems can be simulated by systems of sizes $(1,2,0;1,1,0)$ and $(1,1,0;1,2,0)$, and that these systems can generate all regular languages. Moreover, [9, Theorem 3.3] shows that leftist systems can generate non-context-free languages.

We continue the exploration of the expressive power of leftist systems and show that that adding two-state graph control to insertion-deletion systems of sizes $(1, 2, 0; 1, 1, 0)$ and $(1, 1, 0; 1, 2, 0)$ renders them capable of simulating any 2-tag system (Theorems 4 and 5). That 2-tag systems are universal [3, 22] implies therefore the universality of two-state graph-controlled leftist insertion-deletion systems. Our proofs are based on an extension to insertion and deletion rules we introduce in Definition 1, which allows the specification of contexts as regular expressions instead of single fixed words. We show that graph-controlled leftist systems with such extended contexts can simulate any 2-tag system (Theorem 1). Interestingly, in the non-controlled case, allowing extended context in leftist systems does not augment computational power, as normal leftist systems can simulate regular contexts (Theorems 2 and 3). While the same argument does not generally work in the graph-controlled case, it does apply to the construction from the proof of Theorem 1, which leads to the universality result mentioned above.

2 Preliminaries

We do not present here definitions concerning standard concepts of the theory of formal languages and we refer to [27] for more details. We denote by $|w|$ the length of a word w, by $card(A)$ the cardinality of the alphabet A and by REG, CF, CS, and RE the families of regular, context-free, context-sensitive, and recursively enumerable languages, respectively.

An *m-tag system* is the tuple $TS = (m, \mathcal{A}, P)$, where m is a positive integer, $\mathcal{A} = \{a_1, \ldots, a_{n+1}\}$ is a finite alphabet, and P contains rules of the form $a_i \rightarrow \alpha_i$, where $\alpha_i \in A^*$, for $1 \leq i \leq n$. The letter a_{n+1} is called the halting symbol.

A configuration of the tag system TS is a word $w \in \mathcal{A}^*$. The system passes from the configuration $w = a_{i_1} \ldots a_{i_m} w'$, $1 \leq i_j \leq n + 1$, $1 \leq j \leq m$ to the next configuration z by applying one of the productions $a_i \rightarrow \alpha_i$: the first m symbols of w are erased and α_i is added to the end of the word: $w \Longrightarrow z$, if $z = w'\alpha_i$.

A computation of TS over the word $x \in V^*$ is a sequence of configurations $x \Longrightarrow \ldots \Longrightarrow y$, where either $y = a_{n+1}a_{i_1} \ldots a_{i_{m-1}}y'$, or $|y| < m$. In this case we say that TS halts on x and that y is the result of the computation of TS over x, which is denoted by $y = TS(x)$.

Minsky proved that 2-tag systems are universal [3, 22]. Moreover, according to his proof, it is sufficient to consider only tag systems that halt only on the halting symbol and do not have empty productions.

An *insertion-deletion system* is a construct $ID = (V, T, A, I, D)$, where:

- V is an alphabet;
- $T \subseteq V$ is the *terminal* alphabet (the symbols from $V \setminus T$ are called *non-terminals*);
- $A \subseteq V^*$ is the set of *axioms*;
- I, D are finite sets of triples of the form (u, α, v), where u, α ($\alpha \neq \lambda$), and v are strings over V.

The triples in I are *insertion rules*, and those in D are *deletion rules*. An insertion rule $(u, \alpha, v)_{ins} \in I$ indicates that the string α can be inserted between u and v (which corresponds to the rewriting rule $uv \to u\alpha v$), while a deletion rule $(u, \alpha, v)_{del} \in D$ indicates that α can be removed from between the contexts u and v (which corresponds to the rewriting rule $u\alpha v \to uv$). By $\Longrightarrow$ we denote the relation defined by the insertion or deletion rules and by $\Longrightarrow^*$ the reflexive and transitive closure of $\Longrightarrow$.

The language generated by $ID = (V, T, A, I, D)$ is defined by

$$L(ID) = \{w \in T^* \mid x \Longrightarrow^* w \text{ for some } x \in A\}.$$

The complexity of an insertion-deletion system $ID = (V, T, A, I, D)$ is described by the vector $(n, m, m'; p, q, q')$ called *size*, where

$$n = \max\{|\alpha| \mid (u, \alpha, v)_{ins} \in I\}, \quad p = \max\{|\alpha| \mid (u, \alpha, v)_{del} \in D\},$$
$$m = \max\{|u| \mid (u, \alpha, v)_{ins} \in I\}, \quad q = \max\{|u| \mid (u, \alpha, v)_{del} \in D\},$$
$$m' = \max\{|v| \mid (u, \alpha, v)_{ins} \in I\}, \quad q' = \max\{|v| \mid (u, \alpha, v)_{del} \in D\}.$$

We also denote by $INS_n^{m,m'} DEL_p^{q,q'}$ all the languages generated by the families of insertion-deletion systems of size $(n, m, m'; p, q, q')$. Moreover, we define the total size of the system as the sum of all numbers above: $\psi = n + m + m' + p + q + q'$.

If one of the parameters n, m, m', p, q, or q' is not bounded, then we write instead the symbol $*$. If one of the numbers from the pairs m, m' and q, q' is equal to zero (while the other is not), then we say that the corresponding families have a one-sided context. If the $m' = q' = 0$, and $n = p = 1$, the insertion-deletion systems are called *leftist*.

We also recall that the family of insertion-deletion languages of size $(1, 1, 0; 1, 1, 0)$ is incomparable with REG: $(CF \setminus REG) \cap INS_1^{1,0} DEL_0^{0,0} \neq \emptyset$ and $(ba)^+ \notin INS_1^{1,0} DEL_1^{1,0}$ [18].

A *graph-controlled insertion-deletion system* is a construct

$$\Pi = (V, T, A, H, i_0, I_f, R), \text{ where}$$

- V is a finite alphabet,
- $T \subseteq V$ is the *terminal alphabet*,
- $A \subseteq V^*$ is a finite set of *axioms*,
- H is a set of states of Π,
- $i_0 \subseteq H$ is the *initial state*,
- $I_f \subseteq H$ is the set of *final states*, and
- R is a finite set of rules of the form (l, r, E), where r is an insertion or deletion rule over V, $l \in H$, and $E \subseteq H$.

The relation $\{(i, j) \mid (i, r, E) \in R \text{ and } j \in E\}$ defines a graph, called the communication graph of the system. We remark that in the literature the term "graph control" often implies that there is a one-to-one correspondence between the label i and the rule $(i, r, E) \in R$. This corresponds to the point of view that the rules are located on the edges of the communication graph. Another point of view is to place the rules in the nodes of the communication graph. In this paper we do not consider such restrictions, as all corresponding models are equivalent in the computational power and have mostly identical descriptional complexity parameters.

As is common for graph controlled systems, a configuration of Π is represented by a pair (w, i), where $i \in H$ is the current state and w is the current string. A transition $(w, i) \implies (w', j)$ is performed if there is a rule $(i, (u, \alpha, v)_t, E)$ in R such that $w \implies_t w'$ by the insertion/deletion rule $(u, \alpha, v)_t$, $t \in \{ins, del\}$, and $j \in E$. The result of the computation consists of all terminal strings reaching a final state from an axiom and the initial label, i.e.,

$$L(\Pi) = \{w \in T^* \mid (w_0, i_0) \implies^* (w, i_f), \text{ for some } w_0 \in A, \ i_f \in I_f\}.$$

The family of languages generated by graph-controlled insertion-deletion systems having k states and insertion/deletion rules of size $(n, m, m'; p, q, q')$ is denoted as $GC_k INS_n^{m,m'} DEL_p^{q,q'}$. As we deal with universality we define the notion of the *computation* of a system Π on an input word w with respect to a recursive coding ϕ, denoted as $\Pi(\phi(w))$. This can be obtained by replacing the set of axioms A from the definition of Π by $\{\phi(w)\}$ and then by evolving Π as usual, $L(\Pi)$ being considered as the result of the computation.

3   Universality Results

In this section we extend insertion and deletion rules to contexts given by regular expressions rather than by simple strings.

Definition 1. *Given an alphabet V, an extended insertion rule r is the tuple $(E_l, x, E_r)_{ins}$, where $x \in V^*$ and E_l and E_r are regular expressions over V. The rule r can be applied to the string uv to yield uxv, if $u = u_1 u_2$ such that $u_2 \in L(E_l)$ and $v = v_1 v_2$ such that $v_1 \in L(E_r)$. An extended deletion rule is defined in a similar way.*

A (graph-controlled) insertion-deletion system with regular contexts is a (graph-controlled) insertion-deletion system in which extended insertion and deletion rules are allowed.

We will use the same notation for families of languages generated by insertion-deletion systems with regular contexts as for those generated by conventional ones: $INS_n^{m,m'} DEL_p^{q,q'}$, where m, m', q, and q' will be replaced by REG if the corresponding contexts of insertion or deletion rules are allowed to contain regular expressions.

We will now show that any 2-tag system can be simulated by a graph-controlled insertion-deletion system with regular contexts. Next, we show that regular contexts can be reduced to contexts of size $(1, 1, 0; 1, 2, 0)$ or $(1, 2, 0; 1, 1, 0)$.

Theorem 1. *For every tag system TS, there exists a two-state graph-controlled insertion-deletion system Π of size $(1, REG, 0; 1, REG, 0)$ and a recursive coding ϕ such that the following conditions hold:*

- $\Pi(\phi(w)) = \{TS(w)\}$, *if TS halts on w, and*
- $\Pi(\phi(w)) = \emptyset$, *if TS does not halt on w.*

Proof. Consider an arbitrary tag system $TS = (2, \mathcal{A}, P)$. We will now construct the extended graph-controlled insertion-deletion system $\Pi(\beta) = (V, \mathcal{A}, \emptyset, \{1, 2\}, 1, \{1\}, R)$ simulating the computation of TS on the word $\beta \in \mathcal{A}^*$. The alphabet V is defined as follows:

$$V = \{\bar{B}, B, S, F, E, E', Z, Z'\} \cup \{R_i, R_i', R_i'', R_i''', P_i \mid a_i \to \alpha_i \in P\} \cup \mathcal{A}.$$

The coding ϕ is defined as $\phi(\beta) = \bar{B} B \beta S E E' F$.

The simulation of TS happens in several phases; accordingly, we split the rules in R into the following logical groups:

1. generation of the control string of alternating Z and Z', for each $\mathbf{a} \in \mathcal{A}$:

$$
\begin{aligned}
r_{11} &: \left(1, \ (S, \qquad Z, \ \lambda)_{ins}, \ \{1\}\right), \\
r_{12} &: \left(1, \ (S, \qquad Z', \ \lambda)_{ins}, \ \{1\}\right), \\
r_{13} &: \left(1, \ (\mathbf{a}, \qquad S, \ \lambda)_{del}, \ \{1\}\right), \\
r_{14} &: \left(1, \ (\mathbf{a}(ZZ')^*, E, \ \lambda)_{del}, \ \{1\}\right), \\
r_{15} &: \left(1, \ (Z', \qquad E', \lambda)_{del}, \ \{2\}\right);
\end{aligned}
$$

2. deletion of two symbols at the left end of the string and generation of the signal R_i, for each $\mathbf{a} \in \mathcal{A}$:

$$
\begin{aligned}
r_{21} &: \left(2, \ (B, \qquad R_i', \ \lambda)_{ins}, \ \{2\}\right), \\
r_{22} &: \left(2, \ (R_i' a_i, \ R_i'', \ \lambda)_{ins}, \ \{2\}\right), \\
r_{23} &: \left(2, \ (R_i'' \mathbf{a}, \ R_i''', \lambda)_{ins}, \ \{2\}\right), \\
r_{24} &: \left(2, \ (R_i'', \qquad \mathbf{a}, \ \lambda)_{del}, \ \{1\}\right),
\end{aligned}
$$

$$
\begin{aligned}
r_{25} &: \left(1, \ (R_i', \qquad a_i, \ \lambda)_{del}, \ \{1\}\right), \\
r_{26} &: \left(1, \ (R_i', \qquad R_i'', \lambda)_{del}, \ \{1\}\right), \\
r_{27} &: \left(1, \ (B, \qquad R_i', \ \lambda)_{del}, \ \{1\}\right), \\
r_{28} &: \left(1, \ (BR_i''', \ R_i, \ \lambda)_{ins}, \ \{2\}\right),
\end{aligned}
$$

$$
r_{29} : \left(2, \ (B, \qquad R_i''', \lambda)_{del}, \ \{1\}\right);
$$

3. insertion of the right-hand side of a production of TS, for each $\mathbf{a}, \mathbf{b} \in \mathcal{A}$, and $a_i \to \alpha_i \in P$:

$$
\begin{aligned}
r_{31} &: \left(1, \ (R_i \mathbf{a}, \qquad R_i, \ \lambda)_{ins}, \ \{1\}\right), \\
r_{32} &: \left(1, \ (\mathbf{a}, \qquad R_i, \ \lambda)_{del}, \ \{1\}\right), \\
r_{33} &: \left(1, \ (B, \qquad R_i, \ \lambda)_{del}, \ \{1\}\right), \\
r_{34} &: \left(1, \ (B\mathcal{A}^* R_i Z, P_i, \ \lambda)_{ins}, \ \{1\}\right), \\
r_{35} &: \left(1, \ (\mathbf{a} Z P_i, \qquad b, \ \lambda)_{ins}, \ \{1\}\right), \\
r_{36} &: \left(1, \ (\mathbf{a}, \qquad Z, \ \lambda)_{del}, \ \{1\}\right), \\
r_{37} &: \left(1, \ (\mathbf{a} P_i \alpha_i, \qquad Z', \lambda)_{del}, \ \{1\}\right), \\
r_{38} &: \left(1, \ (\mathbf{a}, \qquad P_i, \ \lambda)_{del}, \ \{2\}\right),
\end{aligned}
$$

4. checking of the halting condition and cleanup, where a_{n+1} is the halting symbol of TS:

$$r_{41} : \left(2, \ (\bar{B}, \qquad\quad B, \lambda)_{del}, \ \{1\}\right),$$
$$r_{42} : \left(1, \ (\bar{B}a_{n+1}\mathcal{A}^*, F, \ \lambda)_{del}, \ \{1\}\right),$$
$$r_{43} : \left(1, \ (\emptyset, \qquad\quad \bar{B}, \emptyset)_{del}, \ \{1\}\right).$$

The simulation of the tag system TS by Π is done in 3 stages. During the first stage a repeating sequence of words ZZ' is inserted, their number being equal to the number of steps TS needs to reach the final configuration. This corresponds to the following computation in Π: $(\bar{B}B \, w \, SEE' \, F, 1) \Longrightarrow^* (\bar{B}Bw(ZZ')^k F, 2)$. The second stage repeatedly simulates the application of a production $a_i \to \alpha_i$ by erasing the two starting symbols and by adding the corresponding appendant to the end: $(\bar{B}B a_i a_j w'(ZZ')^k F, 2) \Longrightarrow^* (\bar{B}Bw'\alpha_i(ZZ')^{k-1} F, 2)$. During the last stage the markers $\bar{B}$, B, and F are removed after checking that the first letter of the word is a_{n+1}: $(\bar{B}Ba_{n+1}w'F, 1) \Longrightarrow^* (a_{n+1}w', 1)$.

Now we will discuss each stage in more details. The simulation of the tag system TS starts with the rules of group (1). The symbol S inserts a sequence of the form $(Z|Z')^*$ by rules r_{11} and r_{12}, and is deleted by r_{13}. Then the nonterminal symbol E is deleted permitting to verify that S has inserted an alternating sequence of the form $(ZZ')^*$. Finally, E' is erased by rule r_{15}, moving the system in state 2 and starting the second stage. This sequence of actions corresponds to the following derivation:

$$(\bar{B}B \, wa \, \underline{S}EE' \, F, 1) \overset{r_{11}}{\underset{r_{12}}{\Longrightarrow}}{}^* (\bar{B}B \, w\underline{a} \, S(ZZ')^k EE' \, F, 1)$$
$$\overset{r_{13}}{\Longrightarrow} (\bar{B}B \, wa \, \underline{(ZZ')^k} EE' \, F, 1) \overset{r_{14}}{\Longrightarrow} (\bar{B}B \, wa \, \underline{(ZZ')^{k-1} ZZ'} E' \, F, 1)$$
$$\overset{r_{15}}{\Longrightarrow} (\bar{B}B \, wa \, \underline{S(ZZ')^k} \, F, 2),$$

where $k \in \mathbb{N}$ and underlining indicates the left context of the rule application effecting the transition into the next configuration.

Remark that for the rules from state 1 and from groups (2) and (3) to be applicable, some symbols must be present which may only be inserted in state 2, so if S does not insert the correct alternating sequence, E and E' cannot be erased, and Π blocks on a string with non-terminals. The only exception is r_{36} which can be applied at any time after S is erased, but, as we will see later, if this deletion happens at the incorrect moment, the system will block as well. Rules r_{41} and r_{43}, on the other hand, are also applicable at any moment, but if they are applied too early (while B is still needed), the string will never reach the form required for F to be deleted by r_{42}.

Now we consider the second stage of simulation. During this stage the role of symbols Z and Z' is to ensure that every deletion of the substring $a_i a$ at the beginning of the string is followed by an insertion of the corresponding α_i at the right end of the string, for $a_i \to \alpha_i \in P$. The string should thus contain as many pairs ZZ' as there are steps in a halting computation of TS starting with w. Remark that, after r_{15} is applied, the rules of group (1) can never become applicable again as there are no more necessary symbols.

Whenever Π is in state 2 with a string of the form $\bar{B}Ba_i\mathbf{a}w\,(ZZ')^kF$, rules r_{21} through r_{24} can only be applied, and necessarily in the following order (we only show the evolution of a prefix of $\bar{B}Ba_i\mathbf{a}\,(ZZ')^kF$, because, in state 2, Π cannot change anything outside it):

$$\bar{B}\underline{B}a_i\mathbf{a} \xrightarrow{r_{21}} \bar{B}BR'_i\underline{a_i}\mathbf{a} \xrightarrow{r_{22}} \bar{B}BR'_ia_i\underline{R''_i}\mathbf{a} \xrightarrow{r_{23}} \bar{B}BR'_ia_i\underline{R''_i}\mathbf{a}R'''_i \xrightarrow{r_{24}} \bar{B}BR'_ia_iR''_iR'''_i.$$

The application of r_{24} moves the system back into state 1. The rules of group (3) are not applicable at this moment, because the string contains no instances of R_i yet. The two rules of the second group which can be applied are r_{25} and r_{27}; remark though that applying r_{27} removes R'_i, so r_{26} cannot be applied to erase R''_i anymore. Given that a_i must be erased in order to enable the deletion of R''_i and the insertion of R_i, the following evolution is the only possible one in a terminal derivation (again, we only show the evolution of the prefix):

$$\bar{B}BR'_i\underline{a_i}R''_iR'''_i \xrightarrow{r_{25}} \bar{B}BR'_i\underline{R''_i}R'''_i \xrightarrow{r_{26}} \bar{B}B\underline{R'_i}R'''_i \xrightarrow{r_{27}} \bar{B}\underline{B}R'''_i \xrightarrow{r_{28}} \bar{B}BR'''_iR_i,$$

where the application of r_{28} moves the system back into state 2. This time, however, B is separated from the rest of the string by an instance of R'''_i, so if rule r_{21} is applied instead of r_{29}, neither r_{24} nor r_{29} will ever become applicable, and Π will block in state 2 on a string with non-terminals. Thus the system has to apply r_{29} immediately after the application of r_{28} to arrive in state 1 with the string $\bar{B}BR_iw(ZZ')^kF$, thereby successfully completing the deletion of $a_i\mathbf{a}$ and introducing the corresponding signal symbol R_i.

Rules r_{31}, r_{32}, and r_{33} move the signal R_i to the right end of the string. Remark that if r_{32} and r_{33} erase all the instances of R_i before r_{34} is applied, Π just blocks on a string with non-terminals. On the other hand, the context of r_{34} requires that, for P_i to be inserted, there should be no extra signal symbols in the string; this assures that exactly one insertion happens at the right end of the string per deletion at the left end.

The correct sequence of actions triggered by a signal symbol R_i at the right end of the string is as follows (we only show the evolution of the suffix, because all rules modifying the left end of the string in state 1 require primed R_i symbols):

$$\ldots\mathbf{a}R_iZZ'\,(ZZ')^{k-1}F \xrightarrow{r_{34}} \mathbf{a}R_iZP_iZ'\,(ZZ')^{k-1}F \xrightarrow{r_{32}} \mathbf{a}\underline{Z}P_iZ'\,(ZZ')^{k-1}F$$
$$\xrightarrow{r_{35}}{}^{*} \mathbf{a}\underline{Z}P_i\alpha_iZ'\,(ZZ')^{k-1}F \xrightarrow{r_{36}} \mathbf{a}P_i\underline{\alpha_i}Z'\,(ZZ')^{k-1}F \xrightarrow{r_{37}} \mathbf{a}P_i\alpha_i\,(ZZ')^{k-1}F$$
$$\xrightarrow{r_{38}} \mathbf{a}\alpha_i\,(ZZ')^{k-1}F,$$

where the last derivation step moves the system into state 2 and initiates the next deletion at the left end of the string. Remark that r_{35} is only applicable after R_i has been erased. Furthermore, even though r_{36} may delete Z almost at any moment when Π is in state 1, if this does occur, then both r_{34} and r_{35} are rendered inapplicable, and Π will end up blocking in state 1 on a string with non-terminals. Rule r_{37} can only erase Z' when applications of r_{35} insert the exact substring α_i from the production $a_i \to \alpha_i$. If Z' is not erased, the signal symbol R_j of the following simulation step will not be able to use r_{34} to initiate the insertion of the right-hand side α_j, and Π will block. Finally, the application of r_{38} moves the system into state 2, enabling the next deletion at the left end of the string.

The last stage of the computation is assured by the rules of group (4). Rule r_{41} is applied non-deterministically in order to disable any further deletions and insertions. Then, the end marker F is erased only if the string contains no more service symbols, no more Z or Z', and if the first symbol after $\bar{B}$ is the halting symbol of Π. If these conditions are not met, F will never be erased and Π will block. If F is successfully erased, however, the rule r_{43} is applied removing the last non-terminal symbol and finalizing the simulation of TS.

We now show that, in the case of one-sided systems, regular contexts do not bring additional computational power.

Theorem 2. $INS_1^{REG,0}DEL_1^{REG,0} \subseteq INS_1^{2,0}DEL_1^{1,0}$.

Proof. We give here only the sketch of the proof of the statement which is based on the proof of the result $REG \subsetneq INS_1^{2,0}DEL_1^{1,0}$ from [9].

Any rule $r : (E, x, \lambda)_t$, $t \in \{ins, del\}$, can be simulated as follows. Let $FA = (Q, T, q_0, F, \delta)$ be the finite automaton such that $L(FA) = L(E)$. Consider the following sets of rules:

$$I = \{(a, Q_0, \lambda)_{ins} \mid a \in T\} \cup \{(Q_i a, Q_j, \lambda)_{ins} \mid q_j \in \delta(q_i, a)\}$$
$$\cup \{(Q_f, x, \lambda)_{ins} \mid f \in F, \text{ if } t = ins\},$$
$$D = \{(a, Q_i, \lambda)_{del} \mid a \in T\} \cup \{(Q_f, x, \lambda)_{del} \mid f \in F, \text{ if } t = del\}.$$

We claim that these rules faithfully simulate the action of the extended rule r. The simulation starts by inserting the symbol Q_0 that marks the guess for the leftmost position for the recognition of context E. Then the string is decorated by symbols Q_i according to the transitions of FA. This allows to check if the string to the right of Q_0 belongs to E. In this case a symbol Q_f, $f \in F$ is ultimately inserted into the string. Now this symbol can insert or delete x according to the type t of the rule. Finally, symbols Q_i are cleaned up.

The validity of the simulation is based on the observation that if the full sequence of insertions (checking the contexts) is not performed, then the rule is not applied. Moreover, if the clean-up phase is not completed, then the string will contain non-terminals that will block the corresponding portion of the string from any further evolution.

A similar theorem holds in the case of systems of size $(1, 1, 0; 1, 2, 0)$. It could be immediately deduced from the previous theorem and [9, Lemma 3.3]; we would like to present a simpler construction, however.

Theorem 3. $INS_1^{REG,0}DEL_1^{REG,0} \subseteq INS_1^{1,0}DEL_1^{2,0}$.

Proof. Like for the previous theorem, we shall only give the sketch of the proof.

Any rule $r : (E, x, \lambda)_t, t \in \{ins, del\}$, can be simulated as follows. Let $FA = (Q, T, q_0, F, \delta)$ be the finite automaton such that $L(FA) = L(E)$. Consider the following sets of rules:

$$I = \{(a, Q_i, \lambda)_{ins} \mid a \in T\} \cup \{(Q_f, x, \lambda)_{ins} \mid f \in F, \text{ if } t = ins\},$$
$$D = \{(Q_i a, Q_j, \lambda)_{del} \mid q_j \in \delta(q_i, a)\} \cup \{(a, Q_0, \lambda)_{del} \mid a \in T\}$$
$$\cup \{(Q_f, x, \lambda)_{del} \mid f \in F, \text{ if } t = del\}.$$

We claim that these rules simulate the action of r faithfully. The simulation strategy is a bit different from the proof of Theorem 2. First, a guess about the context is made and the string is decorated by the sequence of symbols Q_i. When the symbol Q_f, $f \in F$, corresponding to final state of E is inserted, an insertion or deletion of x can be performed. Finally, the validity of the context is checked by the deletion rules that require a valid accepting path of FA to be present to the left of Q_f. The difference from Theorem 2 is that at first the symbols Q_i are randomly inserted into the string, and only after that the deletion rules check that these symbols were inserted in the correct order. In particular, this means that the insertion or deletion of x can happen even if the left context does not satisfy E. However, in this case it will be impossible to erase the remaining non-terminals Q_i.

While insertion-deletion systems of sizes $(1, 2, 0; 1, 1, 0)$ and $(1, 1, 0; 1, 2, 0)$ can simulate any extended insertion-deletion system of size $(1, REG, 0; 1, REG, 0)$, the same statement cannot be directly extended to the graph-controlled case. Indeed, the simulation of extended contexts is based on inserting additional symbols; repeating the same approach for graph-controlled systems would make it possible to switch states right in the middle of the verification of a context of an extended rule allowing some incorrect behavior. For example consider a system containing the following three rules:

$$1 : \big(1, (ab, x, \lambda)_{ins}, \{2\}\big), \quad 2 : \big(2, (d, a, \lambda)_{ins}, \{1\}\big), \quad 3 : \big(1, (a, a, \lambda)_{ins}, \{2\}\big),$$

and suppose that we use the approach from the proof of Theorem 3 to simulate rule 1 with rules of size $(1, 1, 0; 1, 2, 0)$. We will use the following rules to insert state symbols:

$$1a : \big(1, \ (b, Q_2, \lambda)_{ins}, \ \{1\}\big), \quad 1b : \big(1, \ (a, \ Q_1, \lambda)_{ins}, \ \{1\}\big),$$
$$1c : \big(1, \ (a, Q_0, \lambda)_{ins}, \ \{1\}\big), \quad 1d : \big(1, \ (Q_2, x, \ \lambda)_{ins}, \ \{2\}\big),$$

for all symbols a, and the following deletion rules to attempt to verify the context of rule 1:

$$1e : \big(2, \ (Q_1 b, Q_2, \lambda)_{del}, \ \{2\}\big), \quad 1f : \big(2, \ (Q_0 a, Q_1, \lambda)_{del}, \ \{2\}\big),$$
$$1g : \big(2, \ (a, \quad Q_0, \lambda)_{del}, \ \{2\}\big),$$

for all symbols a. In the configuration $(db, 1)$ of the original system, no rule is applicable. However, in the new system with rules of size $(1, 1, 0; 1, 2, 0)$, the following sequence of rule applications is possible:

$$(d\underline{b}, 1) \xRightarrow{1a} (dbQ_2, 1) \xRightarrow{1d} (\underline{d}bQ_2x, 2) \xRightarrow{2} (d\underline{a}bQ_2x, 1) \xRightarrow{3} (\underline{d}aabQ_2x, 2)$$
$$\xRightarrow{2} (daa\underline{a}bQ_2x, 1) \xRightarrow{1b} (da\underline{a}aQ_1bQ_2x, 1) \xRightarrow{1c} (daaQ_0a\underline{Q_1}bQ_2x, 1)$$
$$\xRightarrow{1e} (daa\underline{Q_0a}Q_1bx, 1) \xRightarrow{1f} (da\underline{a}Q_0abx, 1) \xRightarrow{1g} (daaabx, 1).$$

Thus, even though the initial string only partially matches the context ab, by switching to state 2 and after that back to state 1, the missing a is inserted and the context is successfully validated. Remark that the state switch should be done on the insertion of x by rule $1d$, because otherwise several occurrences of x can be introduced into the string. In a more general manner, because states are also in play, it may not be possible to reorder the derivation in such a way that the string fully corresponds to the contexts of the simulated rule r at one given moment, which means that a "simulation" of r may be successfully completed even in the situations in which r itself could never be applied.

The above issue does not occur when simulating regular contexts with systems of size $(1, 2, 0; 1, 1, 0)$, because the insertion or deletion of x is done after all of the state symbols checking the context have been inserted, from left to right (cf. proof of the Theorem 2). However, such systems have another problem – symbol Q_f may not necessarily be deleted immediately; then the insertion or deletion of x can happen twice, even if the left context was changed to not match the rule anymore.

Yet, simulation of regular contexts by rules of size $(1, 2, 0; 1, 1, 0)$ is still possible for the construction from Theorem 1, because the situation we have just described cannot happen. Indeed, when the symbol E is erased by a simulation of r_{14}, for example, the system has already assured the correct form of the string to the left of it. Since moving into state 2 is only possible by r_{15} at this time, we are also sure that the string does change after the symbol-by-symbol checking of the context of r_{14} verifies that S is no longer present.

A slightly more complex analysis is needed for the rules of group (3). When the verification of the context of r_{34} is finished, we know that the string contained $B\mathcal{A}^*R_iZ$ some steps ago, but r_{32} and r_{36} might have erased R_i and Z in the meantime, so r_{38} could be applied thereby allowing one more deletion of two symbols at the end of the string. Note, however, that Z' would not be erased, so the next signal symbol R_j would not be able to trigger an insertion of P_j and the system would block. A similar argument is valid for r_{35}: the Z to the left of P_i should stay in the string in order for all of the symbols of α_i to be inserted, or else Z' will not be deleted. In the case of r_{37}, again, if P_i is deleted before Z', the system blocks.

Finally, when the context of rule r_{42} is completely matched, the only modification that may happen to the string before F is erased is the deletion of O by r_{43}, but this behaviour does not break the simulation of the tag system. Hence, we obtain the following statement.

Theorem 4. *For every tag system TS there exists a two-state graph-controlled insertion-deletion system Π of size $(1, 2, 0; 1, 1, 0)$, and a recursive coding ϕ such that the following conditions hold:*

- $\Pi(\phi(w)) = \{TS(w)\}$, if TS halts on w, and
- $\Pi(\phi(w)) = \emptyset$, if TS does not halt on w.

A symmetric statement for the case of graph-controlled insertion-deletion systems of size $(1,1,0;1,2,0)$ is also true, but the simulation of the construction from Theorem 1 is less straightforward than for systems of size $(1,2,0;1,1,0)$, because, in the case of deletion rules with two-symbol contexts, the simulation of regular rules starts by an insertion of a state symbol at the rightmost end of the substring to be matched. It is therefore possible that the action of a rule is produced before its context is verified, as we have seen above. We will now analyze those rules of the construction from Theorem 1 which are not of the size $(1,1,0;1,2,0)$ one by one, and describe how they can be correctly simulated.

To deal with r_{14}, we will simulate such a finite automaton corresponding to the expression $\mathbf{a}(ZZ')^*$ in which the first state is only visited once, in the initial configuration of the automaton. We will then introduce the symbol $Q_0^{(14)}$, representing this state, into the axiom, before S, giving $\bar{B}B\beta Q_0^{(14)}SEE'F$. The symbol $Q_0^{(14)}$ will be erased by the rule $(1, (\mathbf{a}, Q_0^{(14)}, \lambda)_{del}, 1)$, for all $\mathbf{a} \in \mathcal{A}$. If $Q_0^{(14)}$ is deleted before all state symbols simulating r_{14} are, some of these symbols will stay stuck in the string, because $Q_0^{(14)}$ cannot be inserted. Therefore, the only way to proceed is to erase $Q_0^{(14)}$ after the simulation of r_{14} is successfully finished.

In the case of rules r_{22} and r_{23}, the additional symbols introduced by the simulation will have to be erased before the system moves into state 1, because otherwise they will not be deleted and will block r_{28}, which requires R_i''' to be immediately to the right of B. Rule r_{28} itself will be replaced by the following four rules:

$$\left(1, (R_i''', \quad X_i^{(28)}, \lambda)_{ins}, \{1\}\right), \qquad \left(2, (B, R_i''', \quad \lambda)_{del}, \{2\}\right),$$
$$\left(1, (X_i^{(28)}, R_i, \quad \lambda)_{ins}, \{2\}\right), \qquad \left(2, (B, X_i^{(28)}, \lambda)_{del}, \{1\}\right),$$

where $X_i^{(28)}$ is a new symbol.

For the rest of the rules, usual simulation of regular contexts works correctly. Indeed, a simulation of the rule r_{31} happens in the middle portion of the string, which cannot be altered by rules other than r_{32} or another simulation of r_{31}. In the case of r_{34}, P_i may be inserted even though the string does not have the correct form, moving the system into state 2, and initiating another deletion at the left end. In this situation, however, Z must have been deleted for r_{38} to become applicable, so the string has the form $\bar{B}B\mathcal{A}^*Z'(ZZ')^*F$. Since the next signal symbol R_j cannot interact with Z', this means that the sequence $Z'(ZZ')^*$ will never be deleted. A similar argument is true for r_{35}: if the system switches into state 2 before Z' can be erased, it eventually blocks. As to the simulation of r_{37}, if the system switches away from state 1 before the context is fully verified, the next signal symbol will not be able to insert another P_j, because the state symbols verifying the context of r_{37} will block it on its way to the right end of the string.

Finally, suppose that the simulation of r_{42} erases F at an early stage. Remark that, for this simulation to start at all, F has to be preceded by a symbol from $\mathcal{A}$, which means that the system cannot switch into state 2 while the context of r_{42} is being verified. Therefore, if the string still contains other non-terminals than those simulating r_{24} or $\bar{B}$, the system blocks. Our observations imply the truth of the following statement.

Theorem 5. *For every tag system TS, there exists a two-state graph-controlled insertion-deletion system Π of size $(1, 1, 0; 1, 2, 0)$ and a recursive coding ϕ such that the following conditions hold:*

- $\Pi(\phi(w)) = \{TS(w)\}$, *if TS halts on w, and*
- $\Pi(\phi(w)) = \emptyset$, *if TS does not halt on w.*

4   Conclusions

In this paper, we continued the study of leftist insertion-deletion systems introduced in [9], and showed that systems of sizes $(1, 2, 0; 1, 1, 0)$ and $(1, 1, 0; 1, 2, 0)$ equipped with a two-state graph control mechanism can simulate any 2-tag system, and are therefore universal. This contributes to the study of the computational power of leftist systems started in [9].

The proofs shown in the present work are based on an extension to the conventional insertion and deletion rules, whereby specifying the contexts is done by regular expressions instead of fixed words. We proved that two-state graph-controlled leftist insertion-deletion systems with regular contexts can simulate any 2-tag system.

It turned out that, in the case of leftist insertion-deletion systems without control, considering regular contexts does not increase the expressive power: rules of sizes $(1, 2, 0; 1, 1, 0)$ or $(1, 1, 0; 1, 2, 0)$ can simulate the language of any system of size $(1, REG, 0; 1, REG, 0)$. Even though this statement is not generally transposable to the graph-controlled case, the specific construction from Theorem 1 can be simulated by conventional leftist rules, which yielded the main result of this paper: two-state graph-controlled insertion-deletion systems of sizes $(1, 2, 0; 1, 1, 0)$ and $(1, 1, 0; 1, 2, 0)$ can simulate any 2-tag system.

An important question left open is whether insertion-deletion systems of sizes $(1, 2, 0; 1, 1, 0)$ or $(1, 1, 0; 1, 2, 0)$ are universal or even computationally complete. We conjecture that this is not the case, because one-symbol one-sided rules can only assure transmission of information in one direction in the string.

The second important open question, which may serve as an intermediate step to solving the previous one, is whether two-state graph-controlled leftist insertion-deletion systems are computationally complete, i.e. whether they can generate all recursively enumerable languages directly, without any coding. Again, our conjecture is negative, because two states only provide a very limited kind of control, which does not seem sufficient for simulating an arbitrary grammar or a Turing machine.

Acknowledgments. The authors would like to acknowledge the support of ANR project SynBioTIC.

References

1. Benne, R.: RNA Editing: The Alteration of Protein Coding Sequences of RNA. Ellis Horwood, Chichester, West Sussex (1993)
2. Biegler, F., Burrell, M.J., Daley, M.: Regulated RNA rewriting: modelling RNA editing with guided insertion. Theoret. Comput. Sci. **387**(2), 103–112 (2007)
3. Cocke, J., Minsky, M.: Universality of tag systems with P = 2. J. ACM **11**(1), 15–20 (1964)
4. Daley, M., Kari, L., Gloor, G., Siromoney, R.: Circular contextual insertions/ deletions with applications to biomolecular computation. In: SPIRE/CRIWG, pp. 47–54 (1999)
5. Freund, R., Kogler, M., Rogozhin, Y., Verlan, S.: Graph-controlled insertion-deletion systems. In: McQuillan, I., Pighizzini, G. (eds.) Proceedings of the Twelfth Annual Workshop on Descriptional Complexity of Formal Systems, vol. 31 of EPTCS, pp. 88–98 (2010)
6. Galiukschov, B.: Semicontextual grammars. Matematicheskaya Logica i Matematicheskaya Lingvistika, pp. 38–50. Tallin University, Russian (1981)
7. Haussler, D.: Insertion and Iterated Insertion as Operations on Formal Languages. PhD thesis, University of Colorado at Boulder (1982)
8. Haussler, D.: Insertion languages. Inf. Sci. **31**(1), 77–89 (1983)
9. Ivanov, S., Verlan, S.: On the lower bounds for leftist insertion-deletion languages. Submitted
10. Ivanov, S., Verlan, S.: Random context and semi-conditional insertion-deletion systems. CoRR, abs/1112.5947 (2011)
11. Ivanov, S., Verlan, S.: About one-sided one-symbol insertion-deletion P systems. In: Alhazov, A., Cojocaru, S., Gheorghe, M., Rogozhin, Y., Rozenberg, G., Salomaa, A. (eds.) CMC 2013. LNCS, vol. 8340, pp. 225–237. Springer, Heidelberg (2014)
12. Kari, L.: On insertion and deletion in formal languages. PhD thesis, University of Turku (1991)
13. Kari, L., Păun, G., Thierrin, G., Yu, S.: At the crossroads of DNA computing and formal languages: characterizing RE using insertion-deletion systems. In: Proceedings of 3rd DIMACS Workshop on DNA Based Computing, pp. 318–333. Philadelphia (1997)
14. Kari, L., Thierrin, G.: Contextual insertions/deletions and computability. Inf. Comput. **131**(1), 47–61 (1996)
15. Kleene, S.C.: Representation of events in nerve nets and finite automata. In: Shannon, C., McCarthy, J. (eds.) Automata Studies, pp. 3–41. Princeton University Press, Princeton, NJ (1956)
16. Krassovitskiy, A.: Complexity and Modeling Power of Insertion-Deletion Systems. PhD thesis, Departament de Filologies Romàniques, Universitat Rovira and Virgili (2011)
17. Krassovitskiy, A., Rogozhin, Y., Verlan, S.: Further results on insertion-deletion systems with one-sided contexts. In: Martín-Vide, C., Otto, F., Fernau, H. (eds.) LATA 2008. LNCS, vol. 5196, pp. 333–344. Springer, Heidelberg (2008)
18. Krassovitskiy, A., Rogozhin, Y., Verlan, S.: Computational power of insertion deletion (P) systems with rules of size two. Nat. Comput. **10**(2), 835–852 (2011)

19. Marcus, S.: Contextual grammars. Revue Roumaine de Mathématiques Pures et Appliquées **14**, 1525–1534 (1969)
20. Margenstern, M., Păun, G., Rogozhin, Y., Verlan, S.: Context-free insertion-deletion systems. Theoret. Comput. Sci. **330**(2), 339–348 (2005)
21. Matveevici, A., Rogozhin, Y., Verlan, S.: Insertion-deletion systems with one-sided contexts. In: Durand-Lose, J., Margenstern, M. (eds.) MCU 2007. LNCS, vol. 4664, pp. 205–217. Springer, Heidelberg (2007)
22. Minsky, M.: Computations: Finite and Infinite Machines. Prentice Hall, Englewood Cliffts, NJ (1967)
23. Petre, I., Verlan, S.: Matrix insertion-deletion systems. Theoret. Comput. Sci. **456**, 80–88 (2012)
24. Păun, G.: Marcus Contextual Grammars. Kluwer Academic Publishers, Norwell, MA, USA (1997)
25. Păun, G., My, N.X.: On the inner contextual grammars. Revue Roumaine de Mathématiques Pures et Appliquées **25**, 641–651 (1980)
26. Păun, G., Rozenberg, G., Salomaa, A.: DNA Computing: New Computing Paradigms. Springer, Heidelberg (1998)
27. Rozenberg, G., Salomaa, A. (eds.): Handbook of Formal Languages. Springer-Verlag, Berlin (1997)
28. Takahara, A., Yokomori, T.: On the computational power of insertion-deletion systems. In: Hagiya, M., Ohuchi, A. (eds.) DNA8 Sapporo. LNCS, vol. 2568, pp. 269–280. Springer, Heidelberg (2002)
29. Verlan, S.: On minimal context-free insertion-deletion systems. J. Automata, Languages Comb. **12**(1–2), 317–328 (2007)
30. Verlan, S.: Study of language-theoretic computational paradigms inspired by biology. Habilitation thesis, Université Paris Est (2010)

Tinput-Driven Pushdown Automata

Martin Kutrib$^{(\boxtimes)}$, Andreas Malcher, and Matthias Wendlandt

Institut für Informatik, Universität Giessen,
Arndtstr. 2, 35392 Giessen, Germany
{kutrib,malcher,matthias.wendlandt}@informatik.uni-giessen.de

Abstract. In input-driven pushdown automata (IDPDA) the input alphabet is divided into three distinct classes and the actions on the pushdown store (push, pop, nothing) are solely governed by the input symbols. Here, this model is extended in such a way that the input of an IDPDA is preprocessed by a deterministic sequential transducer. These automata are called tinput-driven pushdown automata (TDPDA) and it turns out that TDPDAs are more powerful than IDPDAs but still not as powerful as real-time deterministic pushdown automata. Nevertheless, even this stronger model has still good closure and decidability properties. In detail, it is shown that TDPDAs are closed under the Boolean operations union, intersection, and complementation. Furthermore, decidability procedures for the inclusion problem as well as for the questions of whether a given automaton is a TDPDA or an IDPDA are developed. Finally, representation theorems for the context-free languages using IDPDAs and TDPDAs are established.

Keywords: Input driven pushdown automata · Sequential transducers · Real-time deterministic context-free languages · Closure properties · Decidability questions

1 Introduction

In order to describe and to analyze "real-life" problems it is desirable to possess theoretical models which have on the one hand a large expressive power to model a large amount of features of the problems. On the other hand, the models should also be manageable in the sense that the commonly studied decidability issues such as emptiness, inclusion, or equivalence are decidable. With regard to the Chomsky hierarchy, two extremes are the regular languages, represented for example by deterministic or nondeterministic finite automata, and the recursively enumerable languages, represented for example by Turing machines. While the latter class is very powerful and allows to describe almost all practical problems one may think of, it is known owing to the Theorem of Rice that almost nothing is decidable for this class. On the other hand, almost all commonly studied problems are decidable for the former class, but the expressive power of regular languages is often not sufficient. Thus, one has to find an agreement in such a way that the expressive power of a model increases at the expense of losing some decidable properties.

© Springer International Publishing Switzerland 2015
J. Durand-Lose and B. Nagy (Eds.): MCU 2015, LNCS 9288, pp. 94–112, 2015.
DOI: 10.1007/978-3-319-23111-2_7

One such extension are pushdown automata (PDA) which are finite automata enlarged with the storage medium of a pushdown store. An interesting subclass of PDAs is represented by *input-driven* PDAs. The essential idea here is that for such devices the operations on the storage medium are dictated by the input symbols. The first references of input-driven PDAs may be found in [5,14], where input-driven PDAs are introduced as classical PDAs in which the input symbols define whether a push operation, a pop operation, or no operation on the pushdown store has to be performed. The main results obtained there show that the membership problem for input-driven PDAs can be solved in logarithmic space, and that the nondeterministic model can be determinized. More on the membership problem has been shown in [8] where the problem is classified to belong to the parallel complexity class NC^1.

The investigation of input-driven PDAs has been revisited in [1,2], where such devices are called visibly PDA or nested word automata. Some of the results are the classification of the language family described by input-driven PDAs to lie properly in between the regular and the deterministic context-free languages, the investigation of closure properties and decidable questions which turn out to be similar to those of regular languages, and descriptional complexity results for the trade-off occurring when nondeterminism is removed from input-driven PDAs. A recent survey with many valuable references on complexity aspects of input-driven PDAs may be found in [16]. Further aspects such as the minimization of input-driven PDAs and a comparison with other subclasses of deterministic context-free languages have been studied in [6,7] while extensions of the model with respect to multiple pushdown stores or more general auxiliary storages are introduced in [12,13]. Recently, the computational power of input-driven automata using the storage medium of a stack and a queue, respectively, have been investigated in [3,11].

The edge between deterministic context-free languages that are accepted by an IDPDA or not is very small. For example, language $\{\, a^n \$ b^n \mid n \geq 1 \,\}$ is accepted by an IDPDA where an a means a push-operation, b means a pop-operation, and a $\$$ leaves the pushdown store unchanged. On the other hand, the very similar language $\{\, a^n \$ a^n \mid n \geq 1 \,\}$ is not accepted by any IDPDA. Similarly, the language $\{\, w \$ w^R \mid w \in \{a,b\}^+ \,\}$, where w^R denotes the reversal of w, is not accepted by any IDPDA, but if w^R is written down with some marked alphabet $\{\hat{a}, \hat{b}\}$, then language $\{\, w \$ \hat{w}^R \mid w \in \{a,b\}^+ \,\}$ is accepted by an IDPDA. To overcome these obstacles we consider a sequential transducer that translates some input to some output which in turn is the input for an IDPDA. In the above first example such a transducer translates every a before reading $\$$ to a and after reading $\$$ to b. In the second example a and b are translated to a, b or $\hat{a}$, $\hat{b}$, respectively, depending on whether or not $\$$ has been read. We call such a pair of a sequential transducer and an IDPDA *tinput-driven* PDA (TDPDA). To implement the idea without giving the transducers too much power for the overall computation, essentially, we will consider only deterministic injective and length-preserving transducers. The detailed definition of a TDPDA is in Sect. 2. Results on the computational capacity of TDPDAs are obtained in Sect. 3. It turns out that TDPDAs are more powerful than IDPDAs, but less powerful than real-time deterministic pushdown automata. Thus, TDPDAs are a proper generalization of IDPDAs. Moreover, the determinization of TDPDAs is possible for IDPDAs as long as the corresponding sequential transducer is deterministic. IDPDAs have nice closure properties and decidability questions. In Sects. 4 and 5, we show similar results for TDPDAs. In detail, constructions for the closure under the union, intersection, complementation, and inverse homomorphism are given as well as a decidability procedure for inclusion. It should be noted that the constructions are possible as long as the underlying automata have compatible signatures, that is, an identical pushdown behavior on the input symbols. We show that IDPDAs and TDPDAs are *not* closed under union and intersection, and inclusion becomes *undecidable* in case of incompatible signatures. Finally, we present in Sect. 6 a construction that proves that IDPDAs and TDPDAs are sufficient to represent all context-free languages under λ-free homomorphism.

2   Preliminaries

Let Σ^* denote the set of all words over the finite alphabet Σ. The *empty word* is denoted by λ, and $\Sigma^+ = \Sigma^* \setminus \{\lambda\}$. The set of words of length at most $n \geq 0$ is denoted by $\Sigma^{\leq n}$. The *reversal* of a word w is denoted by w^R. For the *length* of w we write $|w|$. We use $\subseteq$ for *inclusions* and $\subset$ for *strict inclusions*.

A classical deterministic pushdown automaton is called input-driven if the next input symbol defines the next action on the pushdown store, that is, pushing a symbol onto the pushdown store, popping a symbol from the pushdown store, or changing the state without modifying the pushdown store. To this end, we assume the input alphabet Σ to be partitioned into the sets Σ_N, Σ_D, and Σ_R, that control the actions state change only (N), push (D), and pop (R). A formal definition is:

Definition 1. *A deterministic input-driven pushdown automaton (IDPDA) is a system $M = \langle Q, \Sigma, \Gamma, q_0, F, \bot, \delta_D, \delta_R, \delta_N \rangle$, where*

1. *Q is the finite set of internal states,*
2. *Σ is the finite set of input symbols partitioned into the sets Σ_D, Σ_R, and Σ_N,*
3. *Γ is the finite set of pushdown symbols,*
4. *$q_0 \in Q$ is the initial state,*
5. *$F \subseteq Q$ is the set of accepting states,*
6. *$\bot \notin \Gamma$ is the empty pushdown symbol,*
7. *δ_D is the partial transition function mapping from $Q \times \Sigma_D \times (\Gamma \cup \{\bot\})$ to $Q \times \Gamma$,*
8. *δ_R is the partial transition function mapping from $Q \times \Sigma_R \times (\Gamma \cup \{\bot\})$ to Q,*
9. *δ_N is the partial transition function mapping from $Q \times \Sigma_N \times (\Gamma \cup \{\bot\})$ to Q.*

A *configuration* of an IDPDA $M = \langle Q, \Sigma, \Gamma, q_0, F, \bot, \delta_D, \delta_R, \delta_N \rangle$ is a triple (q, w, s), where $q \in Q$ is the current state, $w \in \Sigma^*$ is the unread part of the input, and $s \in \Gamma^*$ denotes the current pushdown content, where the leftmost symbol is at the top of the pushdown store. The *initial configuration* for an input string w is set to (q_0, w, λ). During the course of its computation, M runs through a sequence of configurations. One step from a configuration to its successor configuration is denoted by $\vdash$. Let $a \in \Sigma$, $w \in \Sigma^*$, $z, z' \in \Gamma$, and $s \in \Gamma^*$. We set

1. $(q, aw, zs) \vdash (q', w, z'zs)$, if $a \in \Sigma_D$ and $(q', z') \in \delta_D(q, a, z)$,
2. $(q, aw, \lambda) \vdash (q', w, z')$, if $a \in \Sigma_D$ and $(q', z') \in \delta_D(q, a, \perp)$,
3. $(q, aw, zs) \vdash (q', w, s)$, if $a \in \Sigma_R$ and $q' \in \delta_R(q, a, z)$,
4. $(q, aw, \lambda) \vdash (q', w, \lambda)$, if $a \in \Sigma_R$ and $q' \in \delta_R(q, a, \perp)$,
5. $(q, aw, zs) \vdash (q', w, zs)$, if $a \in \Sigma_N$ and $q' \in \delta_N(q, a, z)$,
6. $(q, aw, \lambda) \vdash (q', w, \lambda)$, if $a \in \Sigma_N$ and $q' \in \delta_N(q, a, \perp)$.

So, whenever the pushdown store is empty, the successor configuration is computed by the transition functions with the special empty pushdown symbol $\perp$. As usual, we define the reflexive and transitive closure of $\vdash$ by $\vdash^*$. The language accepted by the IDPDA M is the set $L(M)$ of words for which there exists some computation beginning in the initial configuration and ending in a configuration in which the whole input is read and an accepting state is entered. Formally:

$$L(M) = \{\, w \in \Sigma^* \mid (q_0, w, \lambda) \vdash^* (q, \lambda, s) \text{ with } q \in F, s \in \Gamma^* \,\}.$$

The difference between an IDPDA and a classical deterministic pushdown automaton (DPDA) is that the latter makes no distinction on the types of the input symbols, and may perform λ-moves. However, in all cases, there must not be more than one choice of action for any possible configuration. So, the transition function is defined to be a (partial) mapping from $Q \times (\Sigma \cup \{\lambda\}) \times (\Gamma \cup \{\perp\})$ to $Q \times (\Gamma \cup \{\text{pop}, \text{top}\})$, where it is understood that pop means removing the topmost symbol from the pushdown store, top means letting the content of the pushdown store unchanged, and a symbol of Γ means entering this symbol at the top of the pushdown store. In general, the family of all languages accepted by an automaton of some type X will be denoted by $\mathscr{L}(X)$.

For the definition of tinput-driven pushdown automata we need the notion of *deterministic one-way sequential transducers* (DST) which are basically deterministic finite automata equipped with an initially empty output tape. In every transition a DST appends a string over the output alphabet to the output tape. The transduction defined by a DST is the set of all pairs (w, v), where w is the input and v is the output produced after having read w completely. Formally, a DST is a system $T = \langle Q, \Sigma, \Delta, q_0, \delta \rangle$, where Q is the finite set of internal states, Σ is the finite set of input symbols, Δ is the finite set of output symbols, $q_0 \in Q$ is the initial state, and δ is the partial transition function mapping $Q \times \Sigma$ to $Q \times \Delta^*$. By $T(w) \in \Delta^*$ we denote the output produced by T on input $w \in \Sigma^*$. In the following, we will consider only injective and length-preserving DSTs which are also known as injective Mealy machines. The general definition is given with an eye towards possible extensions of the following model.

Let M be an IDPDA and T be an injective and length-preserving DST. Furthermore, the output alphabet of T is the input alphabet of M. Then, the pair (M, T) is called a *tinput-driven pushdown automaton* (TDPDA) and the language accepted by (M, T) is defined as $L(M, T) = \{ w \in \Sigma^* \mid T(w) \in L(M) \}$.

In order to clarify this notion we continue with an example.

Example 2. Language $L_1 = \{ a^n \$ a^n \mid n \geq 1 \}$ is accepted by a TDPDA. Before reading symbol \$ the transducer maps an a to an a, and after reading \$ it maps an a to a b. Thus, L_1 is translated to $\{ a^n \$ b^n \mid n \geq 1 \}$ which is accepted by some IDPDA.

Similarly, $L_2 = \{ w \$ w^R \mid w \in \{a, b\}^* \}$ can be accepted by some TDPDA. Here, the transducer maps any a, b to a, b before reading \$ and to $\hat{a}, \hat{b}$ after reading \$. This gives the language $\{ w \$ \hat{w}^R \mid w \in \{a, b\}^* \}$ which clearly belongs to $\mathscr{L}$(IDPDA).

Finally, consider $L_3 = \{ a^n b^{2n} \mid n \geq 1 \}$. Here, the transducer maps an a to a and every b alternately to b and c. This gives language $\{ a^n (bc)^n \mid n \geq 1 \}$ which is accepted by some IDPDA: every a implies a push-operation, every b implies a pop, and every c leaves the pushdown store unchanged. ∎

3   Computational Capacity

It is known that the language class accepted by IDPDAs is a proper subset of the deterministic context-free languages [1]. In a TDPDA, the input of the IDPDA is preprocessed by a sequential transducer. We have already seen that TDPDAs are strictly more powerful than IDPDAs. Now the question arises whether a TDPDA can accept languages which are not real-time deterministic context-free. The following theorem answers the question negatively.

Theorem 3. *The family $\mathscr{L}$(TDPDA) is effectively included in the family of real-time deterministic context-free languages.*

Proof. Given a TDPDA (M, T) where $M = \langle Q, \Sigma, \Gamma, q_0, F, \bot, \delta_D, \delta_R, \delta_N \rangle$ is an IDPDA and $T = \langle Q', A, \Sigma, q_0', \delta \rangle$ is an injective length-preserving DST, we will construct a deterministic pushdown automaton $M' = \langle S', A, \Gamma, s_0', F', \bot, \delta' \rangle$ such that $L(M') = L(M, T)$.

The basic idea is that M' first computes the output of the DST T internally and then simulates the IDPDA M. To this end, M' needs to keep track of the states of M and T. Thus, we define $S' = Q \times Q'$ and $s_0' = (q_0, q_0')$. The automaton M' accepts, if the input is read completely and M would be in an accepting state. Hence, $F' = F \times Q'$. The transition function is defined as follows for $p, p' \in Q$, $q, q' \in Q'$, $a \in A$, $a' \in \Sigma$, and $z, z' \in \Gamma$.

$$\delta'((p, q), a, z) = \begin{cases} ((p', q'), \lambda) & \text{if } \delta(q, a) = (q', a') \text{ and } \delta_R(p, a', z) = p', \\ ((p', q'), z) & \text{if } \delta(q, a) = (q', a') \text{ and } \delta_N(p, a', z) = p', \\ ((p', q'), zz') & \text{if } \delta(q, a) = (q', a') \text{ and } \delta_D(p, a', z) = (p', z'). \end{cases}$$

By construction, a word w is accepted by (M, T) if and only if w is accepted by M'. Inspecting δ' shows that M' is indeed a deterministic PDA working in real time. □

The previous theorem gives that the family of languages accepted by tinput-driven automata is a subset of the deterministic context-free languages accepted in real time. The next result shows that this inclusion is proper.

Lemma 4. *The family $\mathscr{L}(TDPDA)$ is a proper subset of the real-time deterministic context-free languages.*

Proof. The language $L = \{\, a^n b^{n+m} a^m \mid n, m \geq 0 \,\}$ is clearly accepted by a deterministic PDA. We will show that L is not accepted by any TDPDA.

In contrast to the assertion, assume that L is accepted by a TDPDA (M, T) with $M = \langle Q, \Sigma, \Gamma, q_0, F, \bot, \delta_D, \delta_R, \delta_N \rangle$ and $T = \langle Q', A, \Sigma, s_0, \delta \rangle$, where T has n states, that is, $|Q'| = n$. Let $w = w'_0 w'_1 w'_2 \in L$ be a word with $w'_0, w'_2 \in \{a\}^*$ and $w'_1 \in \{b\}^*$. Then the output of T on input w is denoted by $w_0 w_1 w_2$ where $|w_i| = |w'_i|$ for $0 \leq i \leq 2$.

When T processes w'_0, it has to enter a cycle after at most n steps. The cycle cannot be left before the first b appears in the input. Similar arguments hold for w'_1 and w'_2. Since T is length-preserving, each w_i, $0 \leq i \leq 2$, has the form

$$y_{i,0} y_{i,1} \cdots y_{i,l_i} (x_{i,0} x_{i,1} \cdots x_{i,m_i})^{t_i} x_{i,0} \cdots x_{i,n_i}, \text{ with } l_i, m_i \leq n, n_i < m_i, t_i \geq 0.$$

Now we turn to the computation of M on $w_0 w_1 w_2$, where the length of each of the three subwords is at least n, and analyze the possible pushdown heights while processing the subwords w_i. Since the lengths of the initial parts $y_{i,0} y_{i,1} \cdots y_{i,l_i}$ are at most n, the pushdown height after processing it increases or decreases by at most n symbols. In total, during one input cycle $x_{i,0} x_{i,1} \cdots x_{i,m_i}$ automaton M may increase the height of the pushdown store, leave it as it is, or decrease it.

Subword w_0: Assume that, in total, during a cycle of w_0 the pushdown height is not increased. Then the total height of the pushdown store is at most n after processing w_0. Moreover, there are two different prefixes w'_0 and $\hat{w}'_0$ so that M has the same pushdown content and is in the same state after processing $w_0 = T(w'_0)$ and $\hat{w}_0 = T(\hat{w}'_0)$. Now we can always choose some $w'_1 \in \{b\}^*$ and $w'_2 \in \{a\}^*$ so that $w'_0 w'_1 w'_2$ belongs to L and, thus, is accepted. Since then $\hat{w}'_0 w'_1 w'_2 \notin L$ is accepted as well, we obtain a contradiction and conclude that the total pushdown height is increased during a cycle of w_0.

Subword w_1: Next, we assume that during a cycle of w_1 the pushdown height is decreased. Then we can choose some w'_1 so that $|T(w'_1)| > n \cdot |T(w'_0)|$. So, the height of the pushdown store is at most n after processing $T(w'_0 w'_1)$. Arguing similarly as above, there must be two words w'_1 and $\hat{w}'_1$ so that M has the same pushdown content and is in the same state after processing $w_0 w_1 = T(w'_0 w'_1)$ and $w_0 \hat{w}_1 = T(w'_0 \hat{w}'_1)$. There is a unique $w'_2 \in \{a\}^*$ so that $w'_0 w'_1 w'_2$ belongs to L and, thus, is accepted. Since $w'_0 \hat{w}'_1 w'_2 \notin L$ is accepted as well, we obtain a contradiction and conclude that the total pushdown height is not decreased during a cycle of w_1.

Now, assume that a cycle of w_1 leaves the total pushdown height unchanged. Then, by providing more b's in the input, the cycle can be passed through arbitrarily often. In particular, there must be two words w'_1 and $\hat{w}'_1$ so that M has the same pushdown content and is in the same state after processing the two prefixes $w_0 w_1 = T(w_0' w_1')$ and $w_0 \hat{w}_1 = T(w_0' \hat{w}_1')$. Now the contradiction follows as before. We conclude that the total pushdown height is increased during a cycle of w_1.

Subword w_2: If the pushdown height is not decreased during a cycle of w_2, then its total height is never reduced by more then a constant number of symbols while processing the subword w_2 entirely. Since we know already that the same is true for the subwords w_0 and w_1, the IDPDA M can be simulated by a finite automaton that stores the finite number of accessible symbols at the top of the pushdown store in its state. Since L is not a regular language this is a contradiction. We conclude that the total pushdown height is decreased during a cycle of w_2.

Now we choose two long and different words w_0' and $\hat{w}_0'$ so that M is in the same state and has the same $2n$ symbols on top of the pushdown store after processing $T(w_0')$ and $T(\hat{w}_0')$. The prefix w_0' is completed by w_1' and w_2', where $w_0' w_1' w_2' \in L$ and $|w_2'| < |w_1'|/n$. So, w_2' is such short in comparison to w_1' that the pushdown content pushed while processing $T(w_0')$ is untouched by the computation on $T(w_2')$. It follows that $\hat{w}_0' w_1' w_2' \notin L$ is accepted as well. So, we have a contradiction and obtain that L is not accepted by any TDPDA.   □

Determinization

In the previous part we considered a tinput-driven pushdown automaton as a pair of a deterministic sequential transducer and a deterministic input-driven pushdown automaton. The related model of input-driven automata was also investigated in the nondeterministic case [1]. It is shown there that every nondeterministic input-driven pushdown automaton can be transformed into an equivalent deterministic one.

Now, the question arises whether the nondeterministic version of a tinput-driven pushdown automaton can be determinized as well. There are four different working modes for a tinput-driven pushdown automaton. The sequential transducer can be deterministic or nondeterministic and also the input-driven pushdown automaton may be deterministic or nondeterministic. We use the notation $\text{TDPDA}_{x,y}$ with $x, y \in \{d, n\}$ where x stands for the working mode of the transducer and y for the mode of the input-driven pushdown automaton. For example, $\text{TDPDA}_{n,d}$ is a tinput-driven pushdown automaton with a nondeterministic sequential transducer and a deterministic input-driven pushdown automaton.

Theorem 5. *The family of languages accepted by $\text{TDPDA}_{d,d}$'s is properly included in the family of languages accepted by $\text{TDPDA}_{n,d}$'s.*

Proof. By definition we know that every $\text{TDPDA}_{d,d}$ is in particular a $\text{TDPDA}_{n,d}$.

It remains to be shown that there is a language accepted by a $\text{TDPDA}_{n,d}$, but not by any $\text{TDPDA}_{d,d}$. We will use the language $L = \{\, a^n b^{n+m} a^m \mid n, m \geq 0 \,\}$ from Lemma 4 and prove that L is accepted by a $\text{TDPDA}_{n,d}$ M. This can be done as follows. The nondeterministic sequential transducer writes for every a of the first a-sequence an a as output. Then it writes for every b a b as output until it nondeterministically decides that it has already written as many b's as a's. It continues and writes now for every b an a as output until the first a of the second sequence of a's is reached. Then, for every a a b is output. Subsequently, an IDPDA tests whether its input is of the form $a^n b^n a^m b^m$ for some $m, n \geq 0$. If this is the case, then the input is accepted. Otherwise, it is rejected.

On the other hand, it has been shown in Lemma 4 that L is not accepted by any TDPDA$_{d,d}$.   □

Thus, we can conclude that it is not possible to determinize TDPDA$_{n,d}$'s as well as TDPDA$_{n,n}$'s. It remains for us to consider the determinization of TDPDA$_{d,n}$'s.

Theorem 6. *The family of languages accepted by TDPDA$_{d,n}$'s and TDPDA$_{d,d}$'s coincide.*

Proof. It has been shown in [1] that IDPDAs can be determinized. Applying this construction to the IDPDA belonging to a TDPDA$_{d,n}$, we obtain that every TDPDA$_{d,n}$ can be converted to an equivalent TDPDA$_{d,d}$.   □

4   Closure Properties

In this section, we investigate the closure properties of tinput-driven pushdown automata. For input-driven pushdown automata, strong closure properties have been derived in [1] *provided that* all automata involved share the same partition of the input alphabet. Here we distinguish this important special case from the general one. For easier writing, we call the partition of an input alphabet a *signature*, and say that two signatures $\Sigma = \Sigma_D \cup \Sigma_R \cup \Sigma_N$ and $\Sigma' = \Sigma'_D \cup \Sigma'_R \cup \Sigma'_N$ are *compatible* if and only if

$$\bigcup_{j \in \{D,R,N\}} (\Sigma_j \setminus \Sigma'_j) \cap \Sigma' = \emptyset \quad \text{and} \quad \bigcup_{j \in \{D,R,N\}} (\Sigma'_j \setminus \Sigma_j) \cap \Sigma = \emptyset.$$

We consider first TDPDAs having compatible signatures and identical translations. Later, we will see that IDPDAs and TDPDAs lose some positive closure properties if the signatures are no longer compatible.

Lemma 7. *Let (M, T) and (M', T) be two TDPDAs with compatible signatures. Then TDPDAs accepting the intersection $L(M, T) \cap L(M', T)$, the complement $\overline{L(M, T)}$, and the union $L(M, T) \cup L(M', T)$ can effectively be constructed.*

Proof. Let us first consider the closure under intersection. Since (M, T) and (M', T) have compatible signatures and both TDPDAs apply the sequential transducer T, the closure under intersection follows from the standard construction using the Cartesian product. In detail, we consider the two IDPDAs $M = \langle Q, \Sigma, \Gamma, q_0, F, \bot, \delta_D, \delta_R, \delta_N \rangle$ and $M' = \langle Q', \Sigma', \Gamma', q'_0, F', \bot, \delta'_D, \delta'_R, \delta'_N \rangle$

and define $M'' = \langle Q \times Q', \Sigma \cup \Sigma', \Gamma \times \Gamma', (q_0, q_0'), F \times F', (\bot, \bot), \delta_D'', \delta_R'', \delta_N'' \rangle$ assuming that Σ and Σ' are compatible. The transition functions are defined as follows. Let $q, \hat{q} \in Q$, $q', \hat{q}' \in Q'$, $Z \in \Gamma \cup \{\bot\}$, $\hat{Z} \in \Gamma$, $Z' \in \Gamma' \cup \{\bot\}$, and $\hat{Z}' \in \Gamma'$. For $a \in \Sigma_D \cap \Sigma_D'$, we define $\delta_D''((q, q'), a, (Z, Z')) = ((\hat{q}, \hat{q}'), (\hat{Z}, \hat{Z}'))$ with $\delta_D(q, a, Z) = (\hat{q}, \hat{Z})$ and $\delta_D'(q', a, Z') = (\hat{q}', \hat{Z}')$. For $a \in \Sigma_R \cap \Sigma_R'$, we define $\delta_R''((q, q'), a, (Z, Z')) = ((\hat{q}, \hat{q}'))$ with $\delta_R(q, a, Z) = \hat{q}$ and $\delta_R'(q', a, Z') = \hat{q}'$. For $a \in \Sigma_N \cap \Sigma_N'$, we define $\delta_N''((q, q'), a, (Z, Z')) = ((\hat{q}, \hat{q}'))$ with $\delta_N(q, a, Z) = \hat{q}$ and $\delta_N'(q', a, Z') = \hat{q}'$. For all remaining input symbols $a \in \Sigma \cup \Sigma'$, M'' enters a non-accepting sink state which can never be left once entered. Clearly, (M'', T) is a TDPDA accepting $L(M, T) \cap L(M', T)$.

Next, we consider the closure under complementation. The classical construction for a DPDA is to interchange accepting and non-accepting states. Before doing that two problems have to be overcome. First, the given DPDA may not read its input completely, since some moves are undefined or an infinite λ-loop is entered. Second, it may happen that the given DPDA performs λ-moves leading from an accepting state to a non-accepting state and vice versa. For a TDPDA it is clear from the definition that no λ-moves are performed. Thus, it is sufficient to add a non-accepting state which is entered for so far undefined configurations. This new state cannot be left. It drives the IDPDA component of the TDPDA over the rest of the input obeying the pushdown operations. Finally, accepting and non-accepting states are interchanged.

The effective closure under union follows from the effective closure under intersection and complementation.   □

The next result shows that TDPDAs are closed under inverse homomorphism which is in contrast to IDPDAs.

Lemma 8. *Let (M, T) be a TDPDA and h be a homomorphism. Then a TDPDA accepting $h^{-1}(L(M, T))$ can effectively be constructed.*

Proof. For the construction we will need the following mapping which assigns an integer value to each sequence of input symbols. Let $\Sigma = \Sigma_D \cup \Sigma_R \cup \Sigma_N$ and $\varphi : \Sigma^* \to \mathbb{Z}$ be a mapping such that $\varphi(\lambda) = 0$ and $\varphi(x_1 x_2 \cdots x_n) = \sum_{i=1}^{n} v(x_i)$ setting, for $x \in \Sigma$, $v(x) = 1$ if $x \in \Sigma_D$, $v(x) = -1$ if $x \in \Sigma_R$, and $v(x) = 0$ otherwise.

Now, we will consider an IDPDA $M = \langle Q, \Sigma, \Gamma, q_0, F, \bot, \delta_D, \delta_R, \delta_N \rangle$, an injective and length-preserving DST $T = \langle P, \Delta, \Sigma, \delta, p_0 \rangle$, and an arbitrary homomorphism $h : \Lambda^* \to \Delta^*$. We have to construct a TDPDA (M', T') which accepts $h^{-1}(T(L(M))) = \{ w \in \Lambda^* \mid h(w) \in T(L(M)) \}$. The idea of the construction is first to define T' in such a way that T' simulates T and h in its state set and outputs $T(h(w))$ on given input $w \in \Lambda^*$. Since h may map one symbol to a sequence of symbols, but T' has to be length-preserving, the output alphabet of T' will consist of compressed symbols. In a second step we will construct an IDPDA M' working on an alphabet of compressed symbols and accepting all inputs which are originally and uncompressed accepted by M. Since M' works on compressed input symbols, it will have to work on compressed pushdown symbols as well.

Let $m = \max\{|h(a)| \mid a \in \Lambda\}$ be the maximum length of the image of h. Each (compressed) output symbol will comprise at most m symbols from Δ and two other components ensuring the injectivity of the translation. We now define the DST $T' = \langle P', \Lambda, \Sigma', \delta', p_0' \rangle$ as follows. We set $\Lambda_N = \{a_N \mid a \in \Lambda\}$ and

$$P' = P \times \Sigma^{\leq m-1} \times \{-m+1, -m+2, \ldots, m-1\},$$
$$p_0' = (p_0, \lambda, 0),$$
$$\Sigma' = \Lambda_N \cup \bigcup_{X \in \{D,N,R\}} \left(\Sigma^{\leq m} \times \Lambda \times \Delta^{\leq m}\right)_X.$$

For the transition function δ' we differentiate two cases:

Case 1: If $a \in \Lambda$ such that $h(a) = \lambda$, then we add

$$\delta'((p, d_1 d_2 \cdots d_r, \ell), a) = ((p, d_1 d_2 \cdots d_r, \ell), a_N),$$

for all $p \in P$, $d_1 d_2 \cdots d_r \in \Sigma^{\leq m-1}$, and $-m < \ell < m$. In this case, we just output a symbol a_N which will be ignored by the IDPDA.

Case 2: We have $a \in \Lambda$ such that $h(a) = b_1 b_2 \cdots b_n$ with $n \geq 1$. For $p \in P$, we compute by $\delta(p, b_1 b_2 \cdots b_n) = (p', c_1 c_2 \cdots c_n)$ the state reached and the output produced in T from p on input $b_1 b_2 \cdots b_n$.

To compute the correct index D, R, or N for the output alphabet, we have to check whether the (compressed) symbols $c_1, c_2, \ldots, c_n$ eventually imply a pop-, top-, or push-action in M. To this end, we calculate the value $V = \varphi(c_1 c_2 \cdots c_n)$. If this value is exactly $-m$ or m, we know that a pop- or push-action, respectively, has to take place. If $-m < V < m$, then the pushdown remains unchanged, but V is stored in the state set. If $V < -m$ or $V > m$, then a pop- or push-action, respectively, has to take place, but not all symbols to be output can be compressed into one symbol. Thus, the remaining symbols and their value are stored in the state set. Formally, let $(p, d_1 d_2 \cdots d_r, \ell)$ be a state in P' with $p \in P$, $d_1 d_2 \cdots d_r \in \Sigma^{\leq m-1}$, and $-m < \ell < m$. To compute in T' the next state s and the output o on input a, that is, $\delta'((p, d_1 d_2 \cdots d_r, \ell), a) = (s, o)$, we distinguish five subcases for $K = \ell + \varphi(c_1 c_2 \cdots c_n)$ as follows:

1. If $K = -m$, then $s = (p', \lambda, 0)$ and $o = (d_1 d_2 \cdots d_r c_1 c_2 \cdots c_n, a, h(a))_R$.
2. If $K = m$, then $s = (p', \lambda, 0)$ and $o = (d_1 d_2 \cdots d_r c_1 c_2 \cdots c_n, a, h(a))_D$.
3. If $-m < K < m$, then we define $s = (p', \lambda, \ell + \varphi(c_1 c_2 \cdots c_n))$ and $o = (d_1 d_2 \cdots d_r c_1 c_2 \cdots c_n, a, h(a))_N$.
4. If $K < -m$, then we determine the maximal integer $1 \leq i \leq n$ such that $\ell + \varphi(c_1 c_2 \cdots c_i) = -m$ and we set $s = (p', c_{i+1} c_{i+2} \cdots c_n, \varphi(c_{i+1} c_{i+2} \cdots c_n))$ and $o = (d_1 d_2 \cdots d_r c_1 c_2 \cdots c_i, a, h(a))_R$.
5. If $K > m$, then we determine the maximal integer $1 \leq i \leq n$ such that $\ell + \varphi(c_1 c_2 \cdots c_i) = m$ and we set $s = (p', c_{i+1} c_{i+2} \cdots c_n, \varphi(c_{i+1} c_{i+2} \cdots c_n))$ and $o = (d_1 d_2 \cdots d_r c_1 c_2 \cdots c_i, a, h(a))_D$.

We observe that T' is injective due to the second and third component of the output and length-preserving. On input $w \in \Lambda^*$, T' outputs a compressed version of $T(h(w))$ where up to m symbols are compressed and the index D, R, or N determines the actions on the compressed pushdown of the following IDPDA M' which is defined to work with a compressed pushdown alphabet comprising exactly m symbols. Additionally, the two topmost (compressed) pushdown symbols are simulated in the state set and not in the pushdown store. To realize this, we need an additional dummy pushdown symbol $\perp_D$. Formally, we define $M' = \langle Q', \Sigma', \Gamma', (q_0, \lambda), F', \perp, \delta'_D, \delta'_R, \delta'_N \rangle$ with $\Gamma' = \Gamma^m \cup \{\perp_D\}$, state set $Q' = Q \times (\Gamma^{\leq m-1} \cup \Gamma^{\leq m-1} \times \Gamma^m)$, and $F' = F \times (\Gamma^{\leq m-1} \cup \Gamma^{\leq m-1} \times \Gamma^m)$.

Case 1: We have $a \in \Lambda_N$.

A: Let (q, Z) with $q \in Q$ and $Z \in \Gamma^{\leq m-1}$ be a state in Q'. Then, we set $\delta'_N((q, Z), a, Z') = (q, Z)$ for all $Z' \in \Gamma' \cup \{\perp\}$.

B: Let (q, Z, Z') with $q \in Q$, $Z \in \Gamma^{\leq m-1}$, and $Z' \in \Gamma^m$ be a state in Q'. Then, we set $\delta'_N((q, Z, Z'), a, Z'') = (q, Z, Z')$ for all $Z'' \in \Gamma' \cup \{\perp\}$.

Case 2: We have $(a_1 a_2 \cdots a_n, b, d) \in (\Sigma^{\leq m} \times \Lambda \times \Delta^{\leq m})_X$ with $X \in \{D, N, R\}$.

A: Let $(q, c_1 c_2 \cdots c_r)$ with $q \in Q$ and $c_1 c_2 \cdots c_r \in \Gamma^{\leq m-1}$ be a state in Q'. Consider the computation $(q, a_1 a_2 \cdots a_n, c_1 c_2 \cdots c_r) \vdash^n (q', \lambda, Y_1 Y_2 \cdots Y_k)$ in the IDPDA M with $q' \in Q$ and $Y_i \in \Gamma$ for $1 \leq i \leq k$.

1. If $X = N$, then we know that $k \leq m - 1$ due to the definition of T' and we set $\delta'_N((q, c_1 c_2 \cdots c_r), (a_1 a_2 \cdots a_n, b, d), \perp) = (q', Y_1 Y_2 \cdots Y_k)$.
2. If $X = D$, then we know that $m \leq k \leq m + r$ and we set

$$\delta'_D((q, c_1 c_2 \cdots c_r), (a_1 a_2 \cdots a_n, b, d), \perp) = $$
$$((q', Y_1 Y_2 \cdots Y_{k-m}), Y_{k-m+1} \cdots Y_k), \perp_D).$$

3. The case $X = R$ does not occur, since the pushdown height is less than m.

B: Let $(q, c_1 c_2 \cdots c_r, Z_1 Z_2 \cdots Z_m)$ with $q \in Q$, $c_1 c_2 \cdots c_r \in \Gamma^{\leq m-1}$, and $Z_1 Z_2 \cdots Z_m \in \Gamma^m$ be a state in Q'. Let us consider the following computation in M: $(q, a_1 a_2 \cdots a_n, c_1 c_2 \cdots c_r Z_1 Z_2 \cdots Z_m) \vdash^n (q', \lambda, Y_1 Y_2 \cdots Y_k)$ with $q' \in Q$ and $Y_i \in \Gamma$ for $1 \leq i \leq k$.

1. If $X = N$, then we know that $m \leq k \leq m + r$ and we set, for $Z \in \Gamma'$,

$$\delta'_N((q, c_1 c_2 \cdots c_r, Z_1 Z_2 \cdots Z_m), (a_1 a_2 \cdots a_n, b, d), Z) = $$
$$(q', Y_1 Y_2 \cdots Y_{k-m}, Y_{k-m+1} \cdots Y_k).$$

2. If $X = D$, then we know that $2m \leq k \leq 2m + r$ and we set, for $Z \in \Gamma'$,

$$\delta'_D((q, c_1 c_2 \cdots c_r, Z_1 Z_2 \cdots Z_m), (a_1 a_2 \cdots a_n, b, d), Z) = $$
$$((q', Y_1 Y_2 \cdots Y_{k-2m}, Y_{k-2m+1} \cdots Y_{k-m}), Y_{k-m+1} \cdots Y_k).$$

3. If $X = R$, then we know that $k \leq m - 1$ and we set, for $Z \in \Gamma' \setminus \{\perp_D\}$,

$$\delta'_R((q, c_1 c_2 \cdots c_r, Z_1 Z_2 \cdots Z_m), (a_1 a_2 \cdots a_n, b, d), Z) = $$
$$(q', Y_1 Y_2 \cdots Y_k, Z).$$

For $Z = \perp_D$, we set $\delta'_R((q, c_1 c_2 \cdots c_r, Z_1 Z_2 \cdots Z_m), (a_1 a_2 \cdots a_n, b, d), \perp_D) = (q', Y_1 Y_2 \cdots Y_k)$.

For $w \in \Lambda^*$, M' accepts a compressed version of $T(h(w))$ if and only if $T(h(w))$ is accepted by M. Thus, (M', T') accepts the inverse homomorphic image of $L(M, T)$. □

Next, we turn to non-closure results for TDPDAs even if the signatures are compatible and the transducers are identical.

Lemma 9. $\mathscr{L}(TDPDA)$ *is not closed under concatenation, Kleene star, reversal, and length-preserving homomorphism.*

Proof. Let $L = \{ a^n b^n \mid n \geq 1 \} \cup \{ b^m a^m \mid m \geq 1 \}$. Language L can easily be accepted by a TDPDA (M, T). The sequential transducer T first maps a to a and then b to b if the first symbol read is an a. Otherwise, reading a b first, the transducer maps first b to a and then a to b. In any case, the language output by the transducer is $\{ a^n b^n \mid n \geq 1 \}$ which can be accepted by an IDPDA M. Now, let us consider the concatenation of $L(M, T)$ and assume that a TDPDA for $L(M, T)^2$ can be constructed. Since a TDPDA can simulate a deterministic finite automaton in a second component of its state set, we obtain that $L(M, T)^2 \cap a^+ b^+ a^+ = \{ a^n b^{n+m} a^m \mid n, m \geq 1 \}$ belongs to $\mathscr{L}(TDPDA)$. This is a contradiction to the proof of Lemma 4. Thus, $\mathscr{L}(TDPDA)$ is not closed under concatenation even if the TDPDAs have compatible signatures.

The non-closure under Kleene star and length-preserving homomorphism can be shown similarly observing that $L^* \cap a^+ b^+ a^+ = \{ a^n b^{n+m} a^m \mid n, m \geq 1 \}$ and $h(L') = L$ for $L' = \{ a^n b^n c^m d^m \mid n, m \geq 1 \}$, which is accepted by some IDPDA, and homomorphism h such that $h(a) = h(d) = a$ and $h(b) = h(c) = b$.

Finally, consider language $\{ c^n \$_1 b^m a^n \mid n, m \geq 0 \} \cup \{ d^m \$_2 b^m a^n \mid n, m \geq 0 \}$ which is accepted by some IDPDA. On the other hand, its reversal is not even a real-time deterministic context-free language. □

Remark 10. We would like to remark that the language classes $\mathscr{L}(TDPDA)$ and $\mathscr{L}(IDPDA)$ are not closed under intersection, union, and concatenation in case of incompatible signatures. It suffices to consider the intersection of the languages $\{ a^n b^n c^m \mid n, m \geq 1 \}$ and $\{ a^n b^m c^m \mid n, m \geq 1 \}$ each of which is accepted by some IDPDA. However, the intersection leads to $\{ a^n b^n c^n \mid n \geq 1 \}$ which is not context free. Due to the closure under complementation, both classes cannot be closed under union. For non-closure under concatenation we consider the languages $\{ a^n b^n \mid n \geq 1 \}$ and $\{ b^m a^m \mid m \geq 1 \}$ each of which is accepted by an IDPDA, but their concatenation is not even accepted by any TDPDA due to Lemma 4.

The closure properties discussed in this section are summarized in the following Table 1.

5   Decidability Questions

We recall (see, for example, [10]) that a decidability problem is *semidecidable* (*decidable*) if and only if the set of all instances for which the answer is 'yes' is

Table 1. Closure properties of the language classes discussed. Symbols $\cup_c$, $\cap_c$, and $\cdot_c$ denote union, intersection, and concatenation with compatible signatures. Such operations are not defined for DPDAs and marked with '—'.

	—	$\cup$	$\cap$	$\cup_c$	$\cap_c$	$\cdot$	$\cdot_c$	$*$	$h_{l.p.}$	h^{-1}	REV
DFA	yes	yes	yes	yes	yes	yes	yes	yes	yes	yes	yes
IDPDA	yes	no	no	yes	yes	no	yes	yes	no	no	yes
TDPDA	yes	no	no	yes	yes	no	no	no	no	yes	no
DPDA	yes	no	no	—	—	no	—	no	no	yes	no

recursively enumerable (recursive). Clearly, any decidable problem is also semidecidable, while the converse does not generally hold. An immediate consequence of the effective construction of an equivalent DPDA from a given TDPDA shown in Theorem 3 is the decidability of the decidable problems for deterministic context-free languages. Since an IDPDA is a DPDA by definition, the decidability carries over to the family $\mathscr{L}(\text{IDPDA})$ as well.

Lemma 11. *The problems of equivalence, emptiness, universality, finiteness, infiniteness, and regularity are decidable for TDPDAs and IDPDAs.*

It is known that the inclusion problem for deterministic context-free languages is undecidable. However, for TDPDAs with compatible signatures it *is* decidable.

Theorem 12. *Let (M, T) and (M', T) be two TDPDAs with compatible signatures. Then the inclusion of both TDPDAs is decidable.*

Proof. The inclusion $L(M, T) \subseteq L(M', T)$ can equivalently be expressed by $L(M, T) \cap \overline{L(M', T)} = \emptyset$. Since by Lemma 7 the family $\mathscr{L}(\text{TDPDA})$ is closed under complementation, we obtain that $\overline{L(M', T)}$ is accepted by some TDPDA (M'', T) having the same signature as M'. Since $\mathscr{L}(\text{TDPDA})$ is closed under intersection with compatible signatures by Lemma 7, we obtain a TDPDA (M''', T) which accepts $L(M, T) \cap \overline{L(M', T)}$ and whose emptiness can be tested by Lemma 11. We conclude that the inclusion $L(M, T) \subseteq L(M', T)$ is decidable. □

The role played by the compatibility of the signatures is once more emphasized by the following theorem which states that the inclusion problem becomes even non-semidecidable for incompatible signatures.

The non-semidecidability of the inclusion problem is shown by reduction of the emptiness problem of Turing machines. It is well known that emptiness for such machines is not semidecidable (see, for example, [10]).

In [9] complex Turing machine computations have been encoded in small grammars. Basically, we consider *valid computations of Turing machines*. It suffices to consider deterministic Turing machines with one single tape and one single read-write head. Without loss of generality and for technical reasons, we assume that the Turing machines can halt only after an odd number of moves, accept by halting, make at least three moves, and cannot print a blank. A valid computation is a string built from a sequence of configurations passed through during an accepting computation.

Let Q be the state set of some Turing machine M, where q_0 is the initial state, $T \cap Q = \emptyset$ is the tape alphabet containing the blank symbol, $\Sigma \subset T$ is the input alphabet, and $F \subseteq Q$ is the set of accepting states. Then a configuration of M can be written as a word of the form T^*QT^* such that $t_1 \cdots t_i q t_{i+1} \cdots t_n$ is used to express that M is in state q, scanning tape symbol t_{i+1}, and t_1 to t_n is the support of the tape inscription. For our purpose the valid computations VALC(M) of M are now defined to be the set of strings of the form $\$\bar{w}_1 \$ w_2^R \$ \bar{w}_3 \$ w_4^R \$ \cdots \$ \bar{w}_{2n-1} \$ w_{2n}^R \$$, where $\bar{T}$ and $\bar{Q}$ are disjoint copies of T and Q, $\$ \notin T \cup Q \cup \bar{T} \cup \bar{Q}$, $w_{2i} \in T^*QT^*$ and $w_{2i-1} \in \bar{T}^*\bar{Q}\bar{T}^*$ are configurations of M, $1 \leq i \leq n$, $\bar{w}_1$ is an initial configuration of the form $\bar{q}_0 \bar{\Sigma}^*$, w_{2n} is an accepting configuration of the form T^*FT^*, and w_{i+1} is the successor configuration of w_i, $1 \leq i \leq 2n - 1$.

The valid computations can be decomposed into VALC$_1$(M) which is the set of strings of the form $\$\bar{w}_1 \$ w_2^R \$ \bar{w}_3 \$ w_4^R \$ \cdots \$ \bar{w}_{2n-1} \$ w_{2n}^R \$$, where $\bar{w}_1$ is an initial and w_{2n} is an accepting configuration, and $\bar{w}_{2i+1}$ is the successor configuration of w_{2i}, $1 \leq i \leq n - 1$, and VALC$_2$(M) which is the set of strings of the form $\$\bar{w}_1 \$ w_2^R \$ \bar{w}_3 \$ w_4^R \$ \cdots \$ \bar{w}_{2n-1} \$ w_{2n}^R \$$, where $\bar{w}_1$ is an initial and w_{2n} is an accepting configuration, and w_{2i} is the successor configuration of $\bar{w}_{2i-1}$, $1 \leq i \leq n$. Clearly, the intersection VALC$_1$(M) $\cap$ VALC$_2$(M) is exactly VALC(M). The next lemma gives a construction of an IDPDA accepting VALC(M).

Lemma 13. *Let M be a Turing machine. Then IDPDAs accepting VALC$_1$(M) and VALC$_2$(M) can effectively be constructed from M.*

Proof. The IDPDA M_1 accepting VALC$_1$(M) uses the input symbols $\Sigma_N = \{\$\}$, $\Sigma_D = \bar{Q} \cup \bar{T}$, $\Sigma_R = Q \cup T$. Whenever it starts to read a configuration with odd number, it pushes all symbols read. In addition it remembers the last three symbols read in its finite control until the state symbol of that configuration is the middle one of these three. When the $\$$ appears in the input, M_1 changes its mode. Now it pops a symbol for every symbol read, thus, verifying that the current configuration is the reversal of the successor configuration of the previous one. Both configurations differ only locally at the state symbol. But from the information remembered in the finite control, the differences can be computed and verified. In addition M_1 checks in its finite control whether $\bar{w}_1$ is an initial configuration, and whether the last configuration is an accepting one.

The IDPDA M_2 accepting VALC$_2$(M) works similarly. It uses the input symbols $\Sigma_N = \{\$\}$, $\Sigma_D = Q \cup T$, and $\Sigma_R = \bar{Q} \cup \bar{T}$ instead. In addition it just reads $\bar{w}_1$ (popping from the empty pushdown). □

Now we are prepared to prove the undecidability of the inclusion. Since it is shown for IDPDAs, the result carries over to TDPDAs even if the associated transducers are the same.

Theorem 14. *Let (M,T) and (M',T) be two TDPDAs. Then the inclusion $L(M,T) \subseteq L(M',T)$ is not semidecidable. Let M and M' be two IDPDAs. Then the inclusion $L(M) \subseteq L(M')$ is not semidecidable.*

Proof. We have to show the assertion for IDPDAs only, since IDPDAs are particular TDPDAs. Let M be a Turing machine. From M the two IDPDAs M_1 and M_2 accepting $\text{VALC}_1(M)$ and $\text{VALC}_2(M)$ are constructed according to Lemma 13. Since the family $\mathscr{L}(\text{IDPDA})$ is closed under complementation, an IDPDA M' accepting $\overline{L(M_2)}$ can be constructed.

In contrast to the assertion, assume that the inclusion problem is semidecidable. Then the inclusion $L(M_1) \subseteq L(M') = \overline{L(M_2)}$ is semidecidable. This is equivalent to semidecide $L(M_1) \cap L(M_2) = \emptyset$ which implies that the emptiness of $\text{VALC}(M)$, and hence of $L(M)$, is semidecidable. This is a contradiction.   □

We conclude this section with another decidability problem. Given a deterministic pushdown automaton M and a sequential transducer T, is (M,T) a TDPDA or not? Essentially, this question reduces to the question of whether M is an IDPDA or not. If the output alphabet of T is equal to the input alphabet of M and T is injective and length-preserving, then (M,T) is a TDPDA if and only if M is an IDPDA.

First we present an algorithm which tests whether a given DPDA is an IDPDA.

Theorem 15. *Let M be a DPDA. It is decidable whether M is an IDPDA.*

Proof. In order to decide whether a given deterministic pushdown automaton $M = \langle Q, \Sigma, \Gamma, q_0, F, \bot, \delta \rangle$ is input driven, in general, it is not sufficient to inspect the transition function since it may contain surplus transitions for situations that never appear in any computation. These could be transitions with λ-moves or transitions that perform conflicting pushdown operations on the same input symbol.

So, essentially, it remains to be tested whether a transition is applied in some computation or whether it is surplus. To this end, we label the transitions of δ uniquely, say by the set of labels $R = \{r_1, r_2, \ldots, r_m\}$, for some $m \geq 0$. Then we consider words over the alphabet R. On input $u \in R^*$ a DPDA $\tilde{M}$ with all states final tries to imitate a computation of M by applying in every step the transition whose label is currently read. If $\tilde{M}$ accepts some input $u_1 u_2 \cdots u_n$, then there is a computation (not necessarily accepting) of M that uses the transitions $u_1 u_2 \cdots u_n$ in this order. If conversely there is a computation of M that uses the transitions $u_1 u_2 \cdots u_n$ in this order, then $u_1 u_2 \cdots u_n$ is accepted by $\tilde{M}$. So, in order to determine whether a transition with label r_i of M is useful, it suffices to decide whether $\tilde{M}$ accepts an input containing the letter r_i. This decision can be done by testing the emptiness of the deterministic context-free language $L(\tilde{M}) \cap R^* r_i R^*$.

Assume that M' is constructed from M by deleting all surplus transitions. Clearly, M' and M are equivalent. Now, it is checked that there is no transition with a λ-move, and for any input symbol we consider all transitions on this symbol and check whether the pushdown operations are identical. If and only if this is true for all symbols, M is an IDPDA.   □

To decide the general question of whether (M, T) is a TDPDA it is now sufficient to polish the transducer a little bit.

Theorem 16. *Let M be a DPDA and T be a DST. It is decidable whether (M, T) is a TDPDA.*

Proof. By applying Theorem 15 it is first checked that M is an IDPDA. Second, it has to be verified that the output alphabet of T equals the input alphabet of M. Since after the first step all surplus transitions of M are removed, its input alphabet can be determined by inspection of the remaining transitions. Surplus transitions can be removed from T along the lines of the proof of Theorem 15. It should be noted that the emptiness of the output of a DST can be tested the same way as emptiness is tested for deterministic finite automata. After having removed surplus transitions from T, its output alphabet and the question of whether T is length-preserving can be determined by inspection of the remaining transitions. To conclude the proof it suffices to decide the injectivity of T. To this end, we use the result that the functionality of nondeterministic sequential transducers is decidable (see, for example, [17]). Furthermore, it is known (see, for example, [4,18]) that nondeterministic sequential transducers are closed under inversion. To decide the injectivity of T, we construct from T its inverse transducer T^{-1} and test its functionality. Now, T is injective if and only if T^{-1} is functional.   □

The previous decidability problem concerns devices. For the languages represented by the devices, the decidability status is an open problem: Let M be a deterministic pushdown automaton and T be a sequential transducer. Does $L(M, T)$ belong to $\mathscr{L}(\mathrm{TDPDA})$?

6   Representation Theorems

In [15] Myhill has proved that the regular languages are exactly the closure of the finite languages under union, concatenation and iteration. Such results open the possibility to characterize certain language families by, in some sense, simpler ones and some kind of operations. Besides they shed some light on the structure of the family itself that may be used as powerful reduction tool in order to simplify some proofs or constructions.

Here we turn to characterize the context-free languages by the closure of the deterministic (t)input-driven pushdown automata languages under λ-free homomorphism. Thus replacing the nondeterminism and free pushdown operations on the input symbols by λ-free homomorphisms and vice versa.

Theorem 17. *(a) Let L be a language belonging to $\mathscr{L}(IDPDA)$ and h be a λ-free homomorphism. Then $h(L)$ is a context-free language.*

*(b) Let L be a context-free language. Then there exist a λ-free homomorphism h
and an IDPDA M so that L = h(L(M)).*

Proof. (a) Since $\mathscr{L}$(IDPDA) is included in the context-free languages and the
latter are closed under λ-free homomorphisms, the assertion follows immediately.

(b) Let the context-free language L be given as $L(M')$ for some nonde-
terministic pushdown automaton (NPDA) $M' = \langle Q', \Sigma', \Gamma', q_0', F', \bot, \delta' \rangle$, where
the transition function maps $Q' \times \Sigma' \times (\Gamma' \cup \{\bot\})$ to the finite subsets of $Q' \times \Gamma'^*$.
We may assume that M' never pushes more than one symbol and that – except
for pop moves – it never modifies the symbol read at the top of the pushdown
store. Clearly, any NPDA can be transformed into such a normal form. Now, the
transition function δ' can be represented as finite list of transitions of the form
$Q' \times \Sigma' \times (\Gamma' \cup \{\bot\}) \rightarrow Q' \times (\Gamma' \cup \{\text{pop}, \text{top}\})$. We fix an arbitrary list T of
these transitions and number the elements $t_1, t_2, \ldots, t_m$, for some $m \geq 0$.

The IDPDA $M = \langle Q, \Sigma, \Gamma, q_0, F, \bot, \delta_N, \delta_D, \delta_R \rangle$ is defined by $Q = Q'$, $\Gamma = \Gamma'$,
$q_0 = q_0'$, $F = F'$. Furthermore, the input alphabet is given through

$$\Sigma_N = \{\, [a, t, N] \mid a \in \Sigma', t \in T \,\},$$
$$\Sigma_D = \{\, [a, t, D] \mid a \in \Sigma', t \in T \,\}, \text{ and}$$
$$\Sigma_R = \{\, [a, t, R] \mid a \in \Sigma', t \in T \,\}.$$

To conclude the definition of M, for $a \in \Sigma'$, $p, q \in Q$, $z, z' \in \Gamma$, the transition
functions are set as

$$\delta_N(p, [a, t, N], z) = q \qquad \text{if } \delta(p, a, z) = (q, \text{top}) \text{ is transition } t \text{ in } T,$$
$$\delta_D(p, [a, t, D], z) = (q, z') \qquad \text{if } \delta(p, a, z) = (q, z') \text{ is transition } t \text{ in } T, \text{ and}$$
$$\delta_R(p, [a, t, R], z) = q \qquad \text{if } \delta(p, a, z) = (q, \text{pop}) \text{ is transition } t \text{ in } T.$$

The λ-free homomorphism h maps the input triples to their first component,
that is, $h([a, t, S]) = a$, for $a \in \Sigma'$, $t \in T$, and $S \in \{N, D, R\}$.

In order to show that $h(L(M)) = L = L(M')$ we encode accepting computa-
tions of M' as follows. Let $w = a_1 a_2 \cdots a_n \in \Sigma'^*$ be an input from $L(M')$. Then
the set $\varphi(w)$ contains the word

$$[a_1, t_1, S_1][a_2, t_2, S_2] \cdots [a_n, t_n, S_n]$$

if and only if there is an accepting computation of M' so that, for $1 \leq i \leq n$,

$$(q_0', a_1 a_2 \cdots a_n, \lambda) \vdash^* (p, a_i a_{i+1} \cdots a_n, z\gamma) \vdash (q, a_{i+1} \cdots a_n, \gamma_1 \gamma)$$

and $\delta'(p, a_i, z) = (q, op)$ is transition t_i in T and $S_i = N$, $\gamma_1 = z$ if $op = \text{top}$,
$S_i = D$, $\gamma_1 = z'z$ if $op = z' \in \Gamma$, $S_i = R$, $\gamma_1 = \lambda$ if $op = \text{pop}$.

Next we consider the language accepted by M. The idea of the construction
is that M simulates M'. To this end, it gets some information on the transition
chosen by M' as well as on the type of pushdown operation. This information is
provided as second and third component of the input symbols. So, being in some
state p on input symbol $[a, t, S]$, automaton M tries to simulate transition t of M'.

If this transition fits to state p, input symbol a, and the type of the pushdown operation S, then it is simulated by M; otherwise the transition functions δ' are undefined and the simulation blocks rejecting. So, for all $w \in L(M')$ the set $\varphi(w)$ is accepted by M. We conclude $L(M) \supseteq \varphi(L(M'))$.

Now let $w' = [a_1, t_1, S_1][a_2, t_2, S_2] \cdots [a_n, t_n, S_n] \in L(M)$. By construction this implies that M' accepts $w = a_1 a_2 \cdots a_n$ in a computation that uses the sequence of transitions $t_1 t_2 \cdots t_n$. Therefore, $w' \in \varphi(w)$ and, thus, $L(M) \subseteq \varphi(L(M'))$.

Together we have $\varphi(L(M')) = L(M)$. Furthermore, since the homomorphism h simply removes the last two components of the input triple, we obtain $h(\varphi(L(M'))) = L(M')$ and, thus, $h(L(M)) = L(M')$. $\square$

The proof of the previous theorem reveals immediately that the homomorphic characterization of the context-free languages is also by tinput-driven pushdown automata.

Corollary 18. *A language L is context free if and only if there is a λ-free homomorphism h and a TDPDA M so that $L = h(L(M))$.*

7 Conclusion

In this paper, we have introduced a generalization of input-driven automata in such a way that the input is preprocessed by an injective and length-preserving deterministic sequential transducer. We obtained that almost all positive closure and decidability results for IDPDAs with respect to compatible signatures could be carried over to TDPDAs. It would be interesting to know how these results vary when the properties of the underlying transducer are weakened or strengthened. Possible generalizations would be, for example, non-injective or nondeterministic sequential transducers.

References

1. Alur, R., Madhusudan, P.: Visibly pushdown languages. In: Babai, L. (ed.) Symposium on Theory of Computing (STOC 2004), pp. 202–211. ACM (2004)
2. Alur, R., Madhusudan, P.: Adding nesting structure to words. J. ACM **56**, 16 (2009)
3. Bensch, S., Holzer, M., Kutrib, M., Malcher, A.: Input-driven stack automata. In: Baeten, J.C.M., Ball, T., de Boer, F.S. (eds.) TCS 2012. LNCS, vol. 7604, pp. 28–42. Springer, Heidelberg (2012)
4. Bordihn, H., Holzer, M., Kutrib, M.: Economy of description for basic constructions on rational transductions. J. Autom. Lang. Comb. **9**, 175–188 (2004)
5. von Braunmühl, B., Verbeek, R.: Input-driven languages are recognized in $\log n$ space. In: Karpinski, M., van Leeuwen, J. (eds.) Topics in the Theory of Computation, Mathematics Studies, vol. 102, pp. 1–19. North-Holland (1985)
6. Chervet, P., Walukiewicz, I.: Minimizing variants of visibly pushdown automata. In: Kučera, L., Kučera, A. (eds.) MFCS 2007. LNCS, vol. 4708, pp. 135–146. Springer, Heidelberg (2007)

7. Crespi-Reghizzi, S., Mandrioli, D.: Operator precedence and the visibly pushdown property. J. Comput. Syst. Sci. **78**, 1837–1867 (2012)
8. Dymond, P.W.: Input-driven languages are in $\log n$ depth. Inform. Process. Lett. **26**, 247–250 (1988)
9. Hartmanis, J.: Context-free languages and turing machine computations. Proc. Symposia in Applied Mathematics **19**, 42–51 (1967)
10. Hopcroft, J.E., Ullman, J.D.: Introduction to Automata Theory, Languages, and Computation. Addison-Wesley, Reading (1979)
11. Kutrib, M., Malcher, A., Mereghetti, C., Palano, B., Wendlandt, M.: Deterministic input-driven queue automata: finite turns, decidability, and closure properties. Theor. Comput. Sci. **578**, 58–71 (2015)
12. La Torre, S., Madhusudan, P., Parlato, G.: A robust class of context-sensitive languages. In: Logic in Computer Science (LICS 2007), pp. 161–170. IEEE Computer Society (2007)
13. Madhusudan, P., Parlato, G.: The tree width of auxiliary storage. In: Ball, T., Sagiv, M. (eds.) Principles of Programming Languages (POPL 2011), pp. 283–294. ACM (2011)
14. Mehlhorn, K.: Pebbling mountain ranges and its application to DCFL-recognition. In: de Bakker, J.W., van Leeuwen, J. (eds.) Automata, Languages and Programming. LNCS, vol. 85, pp. 422–435. Springer, Heidelberg (1980)
15. Myhill, J.: Finite automata and the representation of events. Technical Report TR 57–624, WADC (1957)
16. Okhotin, A., Salomaa, K.: Complexity of input-driven pushdown automata. SIGACT News **45**, 47–67 (2014)
17. Schützenberger, M.P.: Sur les relations rationnelles. In: Brakhage, H. (ed.) Automata Theory and Formal Languages. LNCS, vol. 33, pp. 209–213. Springer, Heidelberg (1975)
18. Sheng, Y.: Regular languages. In: Rozenberg, G., Salomaa, A. (eds.) Handbook of Formal Languages, vol. 1, pp. 41–110. Springer, Heidelberg (1997)

Reversible Limited Automata

Martin Kutrib$^{(\boxtimes)}$ and Matthias Wendlandt

Institut für Informatik, Universität Giessen,
Arndtstr. 2, 35392 Giessen, Germany
{kutrib,matthias.wendlandt}@informatik.uni-giessen.de

Abstract. A k-limited automaton is a linear bounded automaton that may rewrite each tape square only in the first k visits, where $k \geq 0$ is a fixed constant. It is known that these automata accept context-free languages only. We investigate deterministic k-limited automata towards their ability to perform reversible computations, that is, computations in which every configuration has at most one predecessor. A first result is that, for all $k \geq 0$, sweeping k-limited automata accept regular languages only. In contrast to reversible finite automata, all regular languages are accepted by sweeping 0-limited automata. Then we study the computational power gained in the number k of possible rewrite operations. It is shown that the reversible 2-limited automata accept regular languages only and, thus, are strictly weaker than general 2-limited automata. Furthermore, a proper inclusion between reversible 3-limited and 4-limited automata languages is obtained. The next levels of the hierarchy are separated between every k and $k+3$ rewrite operations. Finally, it turns out that all k-limited automata accept Church-Rosser languages only, that is, the intersection between context-free and Church-Rosser languages contains an infinite hierarchy of language families beyond the deterministic context-free languages.

1   Introduction

Automata working on a tape so that the possible rewrite operations are limited have been studied for a long time. A famous result obtained in [6] considers linear bounded Turing machines. If any tape square may be visited only a constant number of times, it is shown that even linear-time computations cannot accept non-regular languages. This result has been improved to $O(n \log n)$ time in [5]. Recent results [29] show that the upper as well as the lower bound for the size trade-off is double exponential when a machine of this type is converted into a deterministic finite automaton. A generalization of the machines studied in [6] are introduced by Hibbard [7]. He investigated linear bounded automata that may rewrite each tape square only in the first k visits, where k is a fixed constant. However, afterwards the squares can still be visited any number of times (but without rewriting their contents). It is shown in [7] that the nondeterministic variant characterizes the context-free languages provided $k \geq 2$, while there is a tight and strict hierarchy of language classes depending on k for the deterministic variant. One-limited automata, deterministic and nondeterministic, can accept only regular languages.

© Springer International Publishing Switzerland 2015
J. Durand-Lose and B. Nagy (Eds.): MCU 2015, LNCS 9288, pp. 113–128, 2015.
DOI: 10.1007/978-3-319-23111-2_8

Recently, the study of limited automata from the descriptional complexity point of view has been initiated by Pighizzini and Pisoni [26,27]. In [27] it has been shown that the deterministic 2-limited automata characterize the deterministic context-free languages which complements the result on nondeterministic machines. Furthermore, conversions between 2-limited automata and pushdown automata are investigated. For the deterministic case the upper bound for the conversion from 2-limited automata to pushdown automata is double exponential. Conversely the trade-off is shown to be polynomial. Comparisons between 1-limited automata and finite automata are done in [26]. The unary case has recently been studied in [17].

Here we investigate *reversible* k-limited automata. Reversible computations and reversible versions of computational devices have gained much interest in the recent years. For a reversible computational device it is essential that for every configuration which the device may enter there is both a uniquely defined successor and a uniquely defined predecessor configuration. Thus, reversible devices show a forward and backward deterministic behavior. One motivation for studying such devices is given by the physical observation that the loss of information in irreversible computations results in heat dissipation [3,19]. On the other hand, it is also of great theoretical interest how information is processed in computational devices and in which way, if possible, computations can be made information preserving.

Reversible regular languages are studied in [1,8,9,18,28]. However, reversibility is a property of machines and not a property of languages. So, notions as "the family of reversible regular languages" are meaningless unless the reversibility of a regular language is defined by the reversibility of a certain type of device that accepts it. For example, it turned out that reversible *one-way* deterministic finite automata are less powerful than general (possibly irreversible) finite automata. On the other hand, in [9] it has been shown that reversible *two-way* deterministic finite automata characterize the regular languages. Reversible pushdown automata are investigated in [13] where it is shown that their corresponding language family lies properly in between the regular and the deterministic context-free languages. Thus, every regular language can be accepted by some reversible pushdown automaton, but the deterministic context-free language $\{\, a^n b^n \mid n \geq 1 \,\}$ cannot be accepted by any reversible pushdown automaton. Reversibility has been studied also in other computational devices such as space-bounded Turing machines [20], multi-head finite automata [2,14,23,24], queue automata [15] and the massively parallel model of cellular automata [12,22]. Different aspects of reversibility for classical automata are discussed in [10].

Here, we consider the computational capacity of reversible k-limited automata. After a formal definition in the next section, we give meaningful examples. The first shows that the context-free language $\{\, a^n b^n \mid n \geq 1 \,\}$ is accepted by a reversible 4-limited automaton. Later it turns out that 3 rewrite operations are not sufficient and, thus, the two language families are separated. Section 3 is devoted to sweeping k-limited automata. In particular, it is shown that these automata, independently of k, even if they are irreversible accept regular languages only. In contrast to reversible finite automata, all regular languages are accepted already by sweeping 0-limited automata. In Sect. 4 an infinite hierarchy of language classes defined by reversible limited automata is obtained, where the hierarchy depends on the number of rewrite operations allowed. For general 2-limited automata it is known [27] that they characterize the family of deterministic context-free languages. It turns out that *reversible* 2-limited automata are much weaker. In fact, it is shown that they accept regular languages only. Furthermore, the proper inclusion mentioned between reversible 3-limited and 4-limited automata languages is obtained. Higher levels of the hierarchy are separated between every k and $k + 3$ rewrite operations. Finally, in Sect. 5 it turns out that all k-limited automata accept Church-Rosser languages only, that is, the intersection between context-free and Church-Rosser languages contains an infinite hierarchy of language families beyond the deterministic context-free languages.

2   Preliminaries

We write Σ^* for the set of all words over the finite alphabet Σ. The empty word is denoted by λ, and we set $\Sigma^+ = \Sigma^* \setminus \{\lambda\}$. The reversal of a word w is denoted by w^R, and for the length of w we write $|w|$. We use $\subseteq$ for inclusions and $\subset$ for strict inclusions. Let $k \geq 0$ be an integer. A deterministic k-limited automaton is a restricted linear bounded automaton. It consists of a finite state control and a read-write tape whose initial inscription is the input word in between two endmarkers. At the outset of a computation the automaton is in the designated initial state and the head of the tape scans the left endmarker. Dependent on the current state and the currently scanned symbol on the tape the automaton changes its state, rewrites the current symbol on the tape, and moves the head one cell to the left or one cell to the right. However, the rewriting is restricted so that the machine may rewrite each tape square only in the first k visits. Subsequently, the square can still be scanned but the content cannot be changed anymore. So, a deterministic 0-limited automaton is a two-way deterministic finite automaton. An input is accepted if the machine reaches an accepting state and halts.

The original definition of such devices in [7] is based on string rewriting systems whose sentential forms are seen as configurations of automata. Let $u_1 u_2 \cdots u_{i-1} s u_i u_{i+1} \cdots u_n$ be a sentential form that represents the tape inscription $u_1 u_2 \cdots u_n$, the current state s and the head scanning the symbol u_i. Basically, in [7] rewriting rules of the forms $s u_i \rightarrow u_i' s'$, which mean that the state changes from s to s', the current tape square is rewritten from u_i to u_i', and the head is moved to the right, and $u_{i-1} s \rightarrow s' u_{i-1}'$, which mean that the state changes from s to s', the current tape square is rewritten from u_{i-1} to u_{i-1}', and the head is moved to the left, are provided. In this context, an automaton that changes its head direction on a square scans the square twice. In [26,27] and below limited automata are defined in a way that reflects this behavior.

Formally, a *deterministic k-limited automaton* (k-DLA, for short) is a system $M = \langle S, \Sigma, \Gamma, \delta, \triangleright, \triangleleft, s_0, F \rangle$, where S is the finite, nonempty set of *internal* states, Σ is the finite set of *input symbols*, Γ is the finite set of *tape symbols* partitioned into $\Gamma_k \cup \Gamma_{k-1} \cup \cdots \cup \Gamma_0$ where $\Gamma_0 = \Sigma$, $\rhd \notin \Gamma$ is the *left* and $\lhd \notin \Gamma$ is the *right endmarker*, $s_0 \in S$ is the *initial state*, $F \subseteq S$ is the set of *accepting states*, and $\delta : S \times (\Gamma \cup \{\rhd, \lhd\}) \to S \times (\Gamma \cup \{\rhd, \lhd\}) \times \{-1, 1\}$ is the partial transition function, where -1 means to move the head one square to the left, 1 means to move it one square to the right, and whenever $(s', y, d) = \delta(s, \rhd)$ is defined then $y = \rhd$, $d = 1$ and whenever $(s', y, d) = \delta(s, \lhd)$ is defined then $y = \lhd$, $d = -1$.

In order to implement the limited number of rewrite operations, δ is required to satisfy the following condition. For each $(s', y, d) = \delta(s, x)$ with $x \in \Gamma_i$, (1) if $i = k$ then $x = y$, (2) if $i < k$ and $d = 1$ then $y \in \Gamma_j$ with $j = \min\{\lceil \frac{i}{2} \rceil \cdot 2 + 1, k\}$, and (3) if $i < k$ and $d = -1$ then $y \in \Gamma_j$ with $j = \min\{\lceil \frac{i+1}{2} \rceil \cdot 2, k\}$.

It is worth mentioning that these conditions make the a priori global condition of a head turn on some square local. The clever transformation of the original definition to the automata world used in [26,27] gives that, if a square content is from Γ_i then the head position is always to the right of that square if i is odd, and it is to the left of the square if i is even, as long as $i < k$.

A *configuration* of the k-DLA M is a triple (s, v, h), where $s \in S$ is the current state, $v \in \rhd\Gamma^*\lhd$ is the current tape inscription, and $h \in \{0, 1, \ldots, |w| + 1\}$ gives the current head position. If h is 0 the head scans the symbol $\rhd$, if it satisfies $1 \le i \le |w|$, then the head scans the ith letter of w, and if it is $|w| + 1$, then the head scans the symbol $\lhd$. The *initial configuration* for input w is set to $(s_0, \rhd w \lhd, 0)$. During the course of its computation, M runs through a sequence of configurations. One step from a configuration to its successor configuration is denoted by $\vdash$. Let $a_0 = \rhd$ and $a_{n+1} = \lhd$, for $n \ge 0$, then we set $(s, \rhd a_1 a_2 \cdots a_h \cdots a_n \lhd, h) \vdash (s', \rhd a_1 a_2 \cdots a_h' \cdots a_n \lhd, h + d)$ if and only if $(s', a_h', d) = \delta(s, a_h)$.

A k-DLA *halts*, if the transition function is undefined for the current configuration. An input is *accepted* if the automaton halts at some time in an accepting state, otherwise it is rejected.

The *language accepted* by M is $L(M) = \{w \in \Sigma^* \mid w \text{ is accepted by } M\}$.

Now we turn to *reversible* deterministic k-limited automata. Basically, reversibility is meant with respect to the possibility of stepping the computation back and forth. So, the automata have also to be backward deterministic. That is, any configuration must have at most one predecessor which, in addition, is computable by a k-DLA. For reverse computation steps the head of the input tape is always moved to the opposite direction *before* the tape square is read. Therefore, the automaton rereads the input symbol which has been read or written in a preceding forward step.

A k-DLA is said to be *reversible*, abbreviated as k-REVLA, if for any two *distinct* transitions

$$\delta(s, x) = (q, y, d) \quad \text{and}$$
$$\delta(s', x') = (q', y', d'),$$

if $q = q'$, then $d = d'$ and $y \ne y'$.

The first condition means that transitions yielding the same state all have to move the head the same way. The second condition says that for any configuration the predecessor state is uniquely determined by the state (which then implies the head movement) and the work tape symbol read.

A limited automaton is said to be *sweeping* if the direction of the head movement changes only on the endmarkers.

In order to clarify the notions we continue with an example that is later used to separate the computational capacity of 4-REVLA from 3-REVLA.

Example 1. The linear language $\{\, a^n b^n \mid n \geq 1 \,\}$ is accepted by the 4-REVLA $M = \langle S, \Sigma, \Gamma, \delta, \triangleright, \triangleleft, s_0, F \rangle$ with $S = \{s_0, s_1, s_2, s_+, s_a, p_a, s_b, p_b, q_b, r_b\}$, tape symbols $\Gamma_1 = \{a_1\}$, $\Gamma_2 = \{b_2, b_2'\}$, $\Gamma_3 = \{a_3\}$, $\Gamma_4 = \{a_4, b_4, b_4'\}$, and $F = \{s_+\}$.

The principal and natural idea of the construction is that M moves back and forth over the tape whereby in each cycle one a and one b are marked 'already compared'. However, in order to implement this behavior reversibly, we have to overcome some particularities when the symbols are marked.

At the beginning, M moves in state s_0 across the a's and rewrites them to a_1's. When it arrives at the first b, the tape square is marked (to identify it in backward computations).

$$\delta(s_0, \triangleright) = (s_0, \triangleright, 1)$$
$$\delta(s_0, a) = (s_0, a_1, 1)$$
$$\delta(s_0, b) = (p_a, b_2', -1)$$

In the following cycles, M computes tape inscriptions of the form

$$a_1 \cdots a_1 a_3 a_4 \cdots a_4 b_4' b_4 \cdots b_4 b_2 b \cdots b.$$

The states s_a, p_a are used to move the head leftwards across the a_3, a_4, b_4, b_4' symbols and to handle the reversible behavior at the left when another a_1 is compared.

$$\delta(p_a, a_1) = (s_b, a_3, 1) \qquad \delta(s_a, b_4') = (s_a, b_4', -1)$$
$$\delta(s_a, a_4) = (s_a, a_4, -1) \qquad \delta(s_a, a_3) = (p_a, a_4, -1)$$
$$\delta(s_a, b_4) = (s_a, b_4, -1)$$

So, at the end of such a movement the form of the tape inscription is as before, but the a_3 has been rewritten to a_4 and the neighboring a_1 has been rewritten to a_3. Now M is in state s_b. Similarly, the states s_b, p_b, q_b, r_b are used to move the head rightwards across the $a_4, b_4, b_4', b_2, b_2'$ symbols and to handle the reversible behavior at the right when another b is compared.

$$\delta(s_b, a_4) = (s_b, a_4, 1) \qquad \delta(s_b, b_2') = (p_b, b_4', -1)$$
$$\delta(s_b, b_4') = (s_b, b_4', 1) \qquad \delta(s_b, b_2) = (p_b, b_4, -1)$$
$$\delta(s_b, b_4) = (s_b, b_4, 1)$$

After these transitions the tape inscription has the form

$$a_1 \cdots a_1 a_3 a_4 \cdots a_4 b_4' b_4 \cdots b_4 b_4 b \cdots b$$

where the head scans the second b_4 from the right (or a_3 in the first cycle), and M is in state p_b.

$$\delta(p_b, b_4) = (q_b, b_4, 1) \qquad \delta(q_b, b_4) = (r_b, b_4, 1)$$
$$\delta(p_b, a_3) = (q_b, a_3, 1) \qquad \delta(r_b, b) = (s_a, b_2, -1)$$
$$\delta(q_b, b'_4) = (r_b, b'_4, 1)$$

After these transitions, at the end of such rightward movement the form of the tape inscription is as before, but the b_2 has been rewritten to b_4 and the neighboring b has been rewritten to b_2. Now M is in state s_a again.

When all b's are rewritten, M arrives at the right endmarker in state r_b. Now it has to be verified that there are no more a_1's at the left. To this end, the head is moved across the tape in state s_1 until the a_3 at the left is reached. If the following left move places the head on the left endmarker, there are no more symbols a_1 and the input is accepted by a final move in state s_+.

$$\delta(r_b, \lhd) = (s_1, \lhd, -1) \qquad \delta(s_1, a_4) = (s_1, a_4, -1)$$
$$\delta(s_1, b_4) = (s_1, b_4, -1) \qquad \delta(s_1, a_3) = (s_2, a_4, -1)$$
$$\delta(s_1, b'_4) = (s_1, b'_4, -1) \qquad \delta(s_2, \rhd) = (s_+, \rhd, 1)$$

The construction shows that all words from $\{\, a^n b^n \mid n \geq 1 \,\}$ are accepted. If the input does not have the form $a^+ b^+$, the computation blocks rejecting either in state s_0 or r_b. If the number of a's exceeds the number of b's, the computation blocks rejecting in state s_2, and if the number of b's exceeds the number of a's, the computation blocks rejecting in state p_a. So, only words from $\{\, a^n b^n \mid n \geq 1 \,\}$ are accepted. The reversibility of M is verified by an inspection of the transition function. It meets the condition for reversibility. ∎

Example 2. Example 1 can be extended with slight modifications to a 4-REVLA that accepts the still deterministic linear context-free language

$$\{\, xa^\ell \#\# b^m \# c^n \mid x \in \{a, b\}, \ell, m, n \geq 1, \text{ if } x = a \text{ then } \ell = m \text{ else } \ell = n \,\}.$$

To this end, a 4-REVLA M first checks and rewrites the first symbol. For each possibility an extra set of states is used. If $x = a$ the computation is almost the same as in Example 1. Now the initially rewritten a plays the role of the left endmarker and the third # the role of the right endmarker. When at the end of the initial rewriting of the a's to a_1's the head arrives at the first #, in a sequence of alternating right and left moves the two # symbols are rewritten at least four times to some symbol $\#_4$. To this end, different states are used. Subsequently, the head simply passes through these squares. The final sweep of the input beginning in state s_1 is now replaced by a sweep to the right that checks whether there are only c's to the right of the third # and a subsequent sweep to the left that checks whether there are no a_1's left as before.

In the other case, again the initially rewritten b plays the role of the left endmarker. When at the end of the initial rewriting of the a's to a_1's the head arrives at the first #, in a sequence of alternating right and left sweeps across the infix $\#\# b^m \#$ the infix is rewritten to $\#_4 \#_4 b^m_4 \#_4$ and subsequently simply passed through by the head. ∎

3   Sweeping Reversible k-Limited Automata

Limited automata that are restricted to change the direction of the head movement on the endmarkers only, so-called *sweeping* limited automata, are investigated from a descriptional complexity viewpoint in [17]. Since any k-limited automaton accept context-free languages only [7], all unary languages accepted are regular. In [17] the size trade-offs between unary sweeping k-DLA and finite automata are studied. Here we reconsider sweeping k-DLA over arbitrary alphabets and show first that, for any $k \geq 0$, reversible and even irreversible devices accept only regular languages.

Lemma 3. *Let $k \geq 0$, $n \geq 1$, and M be an n-state sweeping k-DLA. Then another sweeping k-DLA accepting $L(M)$ with at most $k + n + 1$ sweeps can effectively be constructed.*

Proof. Any sweeping k-DLA visits every tape square exactly once during each sweep. So, the tape content is fixed after the kth sweep. For the remaining sweeps, if some tape square is entered in the same state twice, automaton M is in a loop and the input cannot be accepted. We conclude that there are at most n further productive sweeps. Now a k-DLA M' is constructed so that it simulates M and counts the number of sweeps up to $k + n$, additionally. If M wants to start the $(k + n + 1)$st sweep, M' enters a non-accepting state, completes the sweep, and halts rejecting at the opposite endmarker. In this way, M' accepts $L(M)$ with at most $k + n + 1$ sweeps.   □

In contrast to several other automata models, Lemma 3 reveals that the restriction to being sweeping is a hard one even for irreversible limited automata.

Theorem 4. *Let $k \geq 0$. The family of languages accepted by sweeping k-DLA is equal to the family of regular languages.*

Proof. Lemma 3 shows that any language accepted by some sweeping k-DLA can be accepted by some k-DLA whose number of sweeps is bounded by a constant. So, any tape square may be visited only a constant number of times. Therefore, the k-DLA works in linear time. It is shown in [6] that even linear-time one-tape one-head Turing machines cannot accept non-regular languages.

Since a sweeping 0-limited automaton can be seen as sweeping two-way finite automaton, every regular language is accepted by some sweeping k-DLA.   □

An immediate consequence is that sweeping *reversible* k-limited automata can accept only regular languages. However, with one sweep and one rewrite operation per square they can accept all regular languages reversibly.

Theorem 5. *Let $k \geq 1$. The family of languages accepted by sweeping k-REVLA is equal to the family of regular languages.*

Proof. Let a regular language $L \subseteq \Sigma^*$ be given through a DFA M with state set S, initial state s_0, accepting states F, and transition function δ.

The apparent idea of the construction of a sweeping reversible 1-limited automaton $M' = \langle S, \Sigma, \Gamma, \delta', \rhd, \lhd, s_0, F \rangle$ accepting L is to simulate M while the sequence of states passed through is written on the tape.

Accordingly, we set $\Gamma_1 = S \times \Sigma$. The 1-REVLA M' starts the computation by moving the head from the left endmarker and keeping state s_0.

$$\delta'(s_0, \rhd) = (s_0, \rhd, 1)$$

Subsequently, M' sweeps over the tape, whereby in each step the current state is written to the current tape square.

$$\delta'(s_i, a) = (s_j, (s_i, a), 1) \text{ if } \delta(s_i, a) = (s_j) \text{ for } s_i, s_j \in S, a \in \Sigma$$

The transition δ' is not defined for the right endmarker. Therefore, the computation necessarily halts after the first sweep. It is accepting if and only if it halts in an accepting state. So, M' accepts L. The reversibility of M' is verified by an inspection of the transition function.   □

Recall that the regular language $a^* b^*$ is not accepted by any reversible one-way DFA, whereas in [9] it has been shown that reversible *two-way* deterministic finite automata characterize the regular languages. This raises the question for the computational capacity of *sweeping* reversible *two-way* deterministic finite automata, that is, the computational capacity of sweeping 0-REVLA.

Theorem 6. *The family of languages accepted by sweeping 0-REVLA is properly included in the family of regular languages.*

Proof. As for one-way DFA, the regular language $a^* b^*$ can be used as witness for the assertion. In any accepting computation, a sweeping 0-REVLA has to change its state in the first sweep (unless the sweep is useless). Moreover, its behavior must become cyclic on the b-block. So, it must have a state with two incoming edges which are labeled by the same input symbol and, thus, cannot be reversible.   □

4   A Hierarchy of Reversible Limited Automata

This section is devoted to an infinite hierarchy of language classes defined by reversible limited automata, where the hierarchy depends on the number of rewrite operations allowed. We know already that the reversible 1-limited automata characterize the regular languages. For general 2-limited automata it is known [27] that they characterize the family of deterministic context-free languages. It turns out that *reversible* 2-limited automata are much weaker.

Theorem 7. *The family of languages accepted by 2-REVLA is equal to the family of regular languages.*

Proof. Let $M = \langle S, \Sigma, \Gamma, \delta, \triangleright, \triangleleft, s_0, F \rangle$ be an arbitrary reversible 2-limited automaton with n states. We consider the computation on an input word w. For any tape square c_j, $1 \leq j \leq |w|$, let t_j be the time step at which c_j is visited the second time (thus, cannot be rewritten anymore). We define two sets of tape squares associated to c_j as follows. Whenever at some time not before t_j the head is moved from c_j to the right and eventually back to c_j without visiting c_j in between, the rightmost square, say c_ℓ, *that is overwritten* during this subcomputation is put into set R_j. If no square is overwritten nothing is put into the set R_j for this subcomputation. In the same way, the set L_j is defined for subcomputations from c_j to the left.

Let $R_j = \{c_{j_1}, c_{j_2}, \ldots, c_{j_r}\}$, $r \geq 0$, where the squares are ordered according to their positions, that is, $c_j < c_{j_1} < c_{j_2} < \cdots < c_{j_r}$ (see Fig. 1). After the subcomputation that returns the head from c_ℓ to c_j all squares in between and including c_j and c_ℓ have been visited at least twice and, thus, their contents are fixed (recall that an automaton that changes its head direction on a square scans the square twice). Denote the time step at which the head changes its direction on c_ℓ by $t(c_\ell)$. So, from the positional order of the squares we derive $t_j < t(c_{j_1}) < t(c_{j_2}) < \cdots < t(c_{j_r})$. Similarly, $L_j = \{c'_{j_1}, c'_{j_2}, \ldots, c'_{j_l}\}$, $l \geq 0$, where $c_j > c'_{j_1} > c'_{j_2} > \cdots > c'_{j_l}$ and $t_j < t(c'_{j_1}) < t(c'_{j_2}) < \cdots < t(c'_{j_l})$.

Now assume that there is at least one cell c_j so that at least one of the sets R_j or L_j, say R_j, contains at least $n+1$ squares. Let s_ℓ be the state in which tape square c_j is reentered after the subcomputation that returns the head from c_ℓ. Since M is reversible, a *reverse* computation starting in state s_ℓ on square c_j would rewrite square c_ℓ to its predecessor inscription, whereby all squares to the right of c_ℓ are not touched. Since $|R_j| \geq n+1$ there are two identical states $s_{j_i} = s_{j_{i'}}$ with $c_{j_i} < c_{j_{i'}}$. But this implies that a *reverse* computation starting in state $s_{j_{i'}}$ on square c_j would rewrite square c_{j_i} to its predecessor inscription whereby square $s_{j_{i'}}$ is not touched. This is a contradiction to the reversibility of M.

We conclude that for all squares c_j both sets R_j and L_j contain at most n elements. Let $(c'_{j_{i'}}, c_{j_i})$ be a pair from $L_j \times R_j$ and let $t(c'_{j_{i'}}) < t(c_{j_i})$. When the head returns to c_j after time step $t(c_{j_i})$ the head can move freely in between the squares $c'_{j_{i'}}$ and c_{j_i} but the tape content in this area is already fixed. So, without leaving the area the head can visit c_j at most n times. Otherwise it would be in an infinite rejecting loop. Leaving the area means to visit another square from L_j or R_j, thus, extending the area, or to touch the endmarker, which does not help. The same argumentation applies if $t(c'_{j_{i'}}) > t(c_{j_i})$. So, the cardinality of L_j and R_j implies that, in any accepting computation, c_j can be visited at most $2 + 2n \cdot n$ times. Since this is true for all squares, M accepts always in linear time. With the result in [6] that one-tape one-head Turing machines working in linear time accept only regular languages it can be deduced that M accepts a regular language. □

The next level of the hierarchy is built by the reversible 3-limited automata. Here we have an open problem. It is not known whether these automata are able to accept some non-regular language. However, there is a deterministic and linear context-free language not accepted by any reversible 3-limited automaton. So,

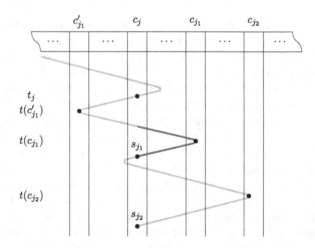

Fig. 1. The head trajectory of an example computation of a 2-REVLA. The states s_{j_1} and s_{j_2} must be different. Otherwise the reverse computation starting in s_{j_1} (red/dark line) applies also for s_{j_2} since the tape content between c_j and c_{j_1} is permanent after $t(c_{j_1})$ (Color figure online).

even with three rewrite operations the unrestricted 2-limited automata cannot be simulated.

Lemma 8. *The real-time deterministic linear language $\{\, a^n b^n \mid n \geq 0 \,\}$ is not accepted by any 3-REVLA.*

Proof. In contrast to the assertion, we assume that there is a 3-REVLA M with state set S that accepts $L = \{\, a^n b^n \mid n \geq 0 \,\}$.

For easier writing we call tape squares that have not been visited three times *writable*.

First we consider accepting computations of M and, in particular, the sub-computations so that the input head is moved from the right half of the tape to the left half and back. We claim that the number of such subcomputations where the head moves on *writable* squares of the left half is not bounded by any constant.

Contrarily assume that there is a constant c so that this number is bounded by c for all inputs $a^n b^n$ with $n \geq 0$. Then, after c subcomputations, the behavior of M on the left half can entirely be described by a table that lists for every state in which M enters the left half what happens. This can either be halting rejecting or accepting or leaving the left half in some state. Since M does not move on writable squares anymore, this behavior cannot change. There are only finitely many of such tables. Furthermore, for any of the first c subcomputations a similar table exists, where these tables may change from subcomputation to subcomputation. However, again there are only finitely many of such tables for all n.

We conclude that there are two numbers n_1 and n_2 so that for the accepting computations on $a^{n_1} b^{n_1}$ and $a^{n_2} b^{n_2}$ the state in which M enters the right half for the first time and all these $c + 1$ tables are the same. Therefore, M accepts $a^{n_1}b^{n_2}$ as well. The contradiction shows the claim.

Now we choose some n large enough so that there are at least $m = |S| + 2$ subcomputations of the accepting computation on $a^n b^n$. Let us denote the sub-computations in chronological order by $C_1, C_2, \ldots, C_m$. If t_i denotes the leftmost writable tape square that is overwritten in the subcomputation C_i, $1 \leq i \leq m$, then we have $t_m < t_{m-1} < \cdots < t_1$. Moreover, at the end of subcomputation C_i all tape squares x of the left half with $x \geq t_i$ have been visited at least three times and, thus, their inscriptions are permanent.

Next, we fix some tape square r that is crossed in all of these subcomputations, say r is the rightmost square of the left half. At latest after C_2 the inscription of r is permanent. Let s_i be the state in which tape square r is entered at the end of subcomputation C_i.

Since M is reversible, a *reverse* computation starting in state s_2 on square r would rewrite square t_2 to its predecessor inscription, whereby the squares t_i with $i \geq 3$ are not touched. Assume now $s_2 = s_i$ for some $3 \leq i \leq m$. Since the tape inscription between t_2 and r does not change after subcomputation C_2, this equality would cause a reverse computation started on r in state s_i to rewrite t_2 whereby the square t_i is not touched. This contradicts the reversibility. So, we conclude that all the states s_i with $i \geq 3$ have to be different from s_2. The same argumentation shows that s_3 has to be different from all the states s_i with $i \geq 4$ and, in general, that s_j has to be different from all states s_i with $i \geq j + 1$. Therefore all states $s_2, \ldots, s_m$ have to be different. Since $m = |S| + 2$ we derive a contradiction to the assumption that M accepts L. $\square$

Example 1 and Lemma 8 separate two levels of the hierarchy. So far, we have

$$\text{REG} = \mathscr{L}(1\text{-REVLA}) = \mathscr{L}(2\text{-REVLA}) \subseteq \mathscr{L}(3\text{-REVLA}) \subset \mathscr{L}(4\text{-REVLA}).$$

In order to obtain an infinite hierarchy of language classes we exploit the informally presented witness languages used in [7] to argue towards an infinite hierarchy of general k-DLA beyond the deterministic context-free languages. We define the languages formally whereby new separator symbols are introduced and sub-words are mirrored which does not change the argumentation in the general case.

For $k \geq 1$, the witness language $L_k \subset \{a, b, c, \#, \$\}^*$ consists of words of the form

$$\$w_1^R\$\$w_3^R\$\$ \cdots \$\$w_{k-1}^R\$\$w_k\$\$ \cdots \$\$w_4\$\$w_2\$x,$$

if k is even, and words of the form

$$\$w_1^R\$\$w_3^R\$\$ \cdots \$\$w_{k-2}^R\$\$w_k^R\$\$w_{k-1}\$\$ \cdots \$\$w_4\$\$w_2\$x,$$

if k is odd, where $x \in \{a, b\}$ and $w_i = a^{\ell_i}\#\#b^{m_i}\#c^{n_i}$, $\ell_i, m_i, n_i \geq 1$, $1 \leq i \leq k$. The words w_i are called subwords in the following.

Whether a word of this form belongs to L_k depends on conditions that are expressed by predicates P, P_{ab} and P_{ac} as follows. In particular, a word of this form belongs to L_k if and only if $P(w_k) = \text{true}$, where

$$P_{ab}(w_i) = \text{true}, \text{ iff } \ell_i = m_i, \qquad P_{ac}(w_i) = \text{true}, \text{ iff } \ell_i = n_i$$

and $P(w_i) =$ true, iff $(P_{ab}(w_i)$ and $P(w_{i-1}))$ or $(P_{ac}(w_i)$ and not $P(w_{i-1}))$ for $i > 1$, with $P(w_1) =$ true, iff $(P_{ab}(w_1)$ and $x = a)$ or $(P_{ac}(w_1)$ and $x = b)$.

Lemma 9. *Let $k \geq 1$. Then language L_k is accepted by some $(k + 4)$-REVLA.*

Proof. In the following we will exploit the constructions from Examples 1 and 2. To this end, assume for a moment that the input word has the correct form. Taking a close look at the constructions in the examples shows that the endmarkers are visited at most twice. Moreover, the constructions can easily be extended such that in case of non-matching numbers of symbols the 4-REVLA does not get stuck somewhere on the input, but in a well-defined state s_- entered only once at the left endmarker. If the original machines get stuck, now a new state is entered that drives the head to the left endmarker. Since all squares that have to be passed through in this situation have not been visited four times, the reversibility is ensured by writing the current state to the current tape square. We mean this modified machine whenever we refer to the examples.

Next we describe the construction of a $(k + 4)$-REVLA M_k accepting L_k. The principal idea is to compute the predicates $P(w_i)$ successively in increasing order. The conditions on the subwords are checked with the machines from the examples, where on the left half of the input mirrored versions of the constructions are used.

In more detail, let w be a given input to M_k. We may consider the tape squares to have $k + 4$ initially blank registers each, that may be filled with information during the visits.

At the beginning, M_k sweeps over the tape whereby it checks the correct format of the input. This can be done by simulating a deterministic finite automaton. The reversibility is ensured by writing the history of the computation into the first registers of the squares. Then the x at the right end is read and remembered in the state while M_k sweeps back to the $\$$ immediately to the right of w_1^R. Since k is a constant this can be done. Again, the reversibility is ensured by writing the history of the computation to the second registers.

Now a mirrored version of the algorithm of Example 2 is applied, where the information from the initial x is already known. For the simulation the four registers $k+1$ to $k+4$ are used. At the end of the algorithm the head of M_k is on the first $\$$ following w_1. It simulates now a state transition to the state s_+ or s_- for the first time. Here we may safely assume that the head is moved to the right since the left step of the original algorithm was just caused by the endmarker. So, the simulation ends on the second $\$$ following w_1, and $P(w_1)$ is computed. The subword w_1 and its immediate surrounding $\$$ are never visited again. Since the endmarkers are visited at most twice in the original algorithm, the reversibility of the constructions in the examples carries over to the reversibility of M_k, since the $\$$ playing the roles of the endmarkers can be rewritten whenever they are visited. Moreover, by the reversibility preserving modifications explained, M_k works reversible so far.

Next, M_k sweeps to the right until it reaches the $\$$ in front of w_2. The reversibility is once again ensured by writing the history of the computation into the third registers of the squares. From here another copy, that means a copy of the state set, of the algorithm of Example 2 is applied, where the information from the initial x is already known from $P(w_1)$. As for the computation on w_1, the four registers $k + 1$ to $k + 4$ are used during the simulation. At the end of the algorithm the head of M_k is on the first \$ in front of w_2. It simulates now a state transition to the (copy of the) state s_+ or s_- for the first time, where the head is moved to the left. So, the simulation ends on the left \$ in front of w_2, and $P(w_2)$ is computed. The subword w_2 and its immediate surrounding \$ are never visited again.

Now, the behavior of M_k continues as before. In the next stage its head sweeps back to the \$ immediately to the right of w_3^R, whereby the reversibility is ensured by writing the history of the computation into the fourth registers of the squares. Again, a new copy of the mirrored version of the algorithm of Example 2 is applied, where the information from the initial x is already known from $P(w_2)$, and so on.

Finally, after having simulated the algorithm from Example 2 on subword w_k, the predicate $P(w_k)$ is known and, correspondingly, M_k can halt accepting or rejecting. Since the subword w_k is checked at latest, it is passed through by previous sweeps most frequently. In particular, this is the very first sweep and one sweep for every subword w_i with $i < k$. So, every cell is visited at most k times. Additionally, 4 visits are used by the algorithm of Example 2. Since M_k is reversible by construction it is in fact a $(k + 4)$-REVLA. □

Now the infinite strict hierarchy follows with the results in [7] where, in essence, it is shown that language L_k is not accepted by any $(k + 1)$-DLA. In fact, our witness languages L_k differ from the original ones by the new separator symbols and by mirroring the subwords on the left. However, thereby the argumentation of [7] is not affected.

Theorem 10. *Let $k \geq 2$. The family of languages accepted by k-REVLA is properly included in the family of $(k + 3)$-REVLA.*

5   Limited Automata and Church-Rosser Languages

In the previous section we derived an infinite hierarchy of reversible k-limited automata. On one hand, all languages accepted by (even nondeterministic) k-limited automata are context free. On the other hand, the hierarchy of reversible k-limited automata includes an infinite language family beyond the deterministic context-free languages. In order to qualify these languages furthermore, we consider the Church-Rosser languages (CRL) that are another popular language family lying properly in between the regular and the context-sensitive languages. Church-Rosser languages have been introduced in [21] via finite, confluent, and length-reducing Thue systems. They are incomparable to the context-free languages [4] and have neat properties. In particular, they contain the deterministic context-free languages as well as their reversals properly [21]. So a natural question is whether they also include the languages accepted by (reversible) k-limited automata. Or with other words, whether the infinite language family beyond the deterministic context-free languages defined by the deterministic k-limited automata belongs still to the intersection of Church-Rosser and context-free languages (see [11] for further results on the intersection).

To answer this question in the affirmative, we exploit the automata characterization of Church-Rosser languages derived in [25], where it is shown that shrinking as well as length-reducing deterministic two-pushdown automata characterize the Church-Rosser languages. This characterization remains valid even for the stateless variant of these automata [16].

A *deterministic two-pushdown automaton* (DTPDA) is a deterministic automaton with two pushdown stores. There is no input tape but the input is provided as the initial contents of the second pushdown store. In general, in every step the DTPDA has access to the topmost k pushdown symbols of both pushdown stores. An input is accepted if and only if the DTPDA halts in an accepting state with empty pushdown stores. Since here we only need DTPDA with $k = 1$, we simplify the following formal definition accordingly. A DTPDA is defined as a system $M = \langle S, \Sigma, \Gamma, \delta, \perp, s_0, F \rangle$, where S is the finite, nonempty set of *internal states*, Σ is the finite set of *input symbols*, $\Gamma \supset \Sigma$ is the finite set of *pushdown symbols*, $s_0 \in S$ is the *initial state*, $\perp \in \Gamma \setminus \Sigma$ is the *bottom marker* of the pushdown stores, $F \subset S$ is the set of *accepting states*, and $\delta : S \times \Gamma \times \Gamma \to Q \times \Gamma^* \times \Gamma^*$ is the partial transition function. In addition, we require that the special symbol $\perp$ can only occur at the bottom of a pushdown store, and that no other letter can occur at that place.

A DTPDA is called *shrinking* if there exists a weight function $\varphi : S \cup \Gamma \to \mathbb{N}$ such that, for all $p \in S$ and all $u, v \in \Gamma$, $(q, u', v') \in \delta(p, u, v)$ implies that $\varphi(q) + \varphi(u') + \varphi(v') < \varphi(p) + \varphi(u) + \varphi(v)$. For a simpler presentation we set $\varphi(\lambda) = 0$.

Theorem 11. *Let $k \geq 0$. The family of languages accepted by k-DLA is properly included in the family of Church-Rosser languages.*

Proof. Let $M = \langle S, \Sigma, \Gamma, \delta, \triangleright, \triangleleft, s_0, F \rangle$ be an arbitrary reversible k-limited automaton. In the following a shrinking DTPDA $M' = \langle S', \Sigma, \Gamma', \delta', \perp, s_0', F' \rangle$ with weight function φ is constructed that accepts $L(M)$.

The principal idea of the simulation is to store the current state and the current tape square content of M in the current state of M'. The left pushdown store of M' contains the tape inscription to the left and the right pushdown store the tape inscription to the right of the current square. In this way, a move of M can be simulated by a move of M' in a straightforward manner. However, the construction is more involved as M' has to be shrinking.

Consider an adjacent block of tape squares that cannot be rewritten anymore. The behavior of M on this block can entirely be described by a table that maps any pair of a state and side of entry to the state and side where the block is left again, or to *accept* or *reject* if the computation is accepting or rejecting within the block. The finite set of such tables derived from M is denoted by T. For convenience the empty word (table) is also included in T. Let $t_v, t_w \in T$ be the two tables for tape inscriptions v and w. If v and w are adjacent, then the tables can easily be merged to the unique table t_{vw}. The empty table is denoted by t_λ. During the simulation of M, tape inscriptions that cannot be rewritten anymore are replaced by such tables. Now it may happen that M' sees tables on top of its pushdown stores that have to be stored as part of its state. Moreover, it may happen that this happens for *both* pushdown stores, so that two tables have to be stored as part of the state. For such a situation we provide states (of M') of the form $(t_u, p, \overrightarrow{t_v})$ which means the simulated state (of M) is p, to the left of the current square is the tape inscription u, followed immediately by the tape inscription v, and M has entered the right block in state p. Here u may be the empty word so that $t_u = t_\lambda$. Symmetrically, we use states of the form $(\overleftarrow{t_u}, p, t_v)$. Finally, $2k + 2$ copies of every table are used to maintain the weights $1, 2, \ldots, 2k + 2$ that φ associates to tables. So, a symbol $t_u^{(i)}$ stands for the table of the tape inscription u so that $\varphi(t_u^{(i)}) = i$, $1 \leq i \leq 2k + 2$. $\square$

References

1. Angluin, D.: Inference of reversible languages. J. ACM **29**, 741–765 (1982)
2. Axelsen, H.B.: Reversible multi-head finite automata characterize reversible logarithmic space. In: Dediu, A.-H., Martín-Vide, C. (eds.) LATA 2012. LNCS, vol. 7183, pp. 95–105. Springer, Heidelberg (2012)
3. Bennett, C.H.: Logical reversibility of computation. IBM J. Res. Dev. **17**, 525–532 (1973)
4. Buntrock, G., Otto, F.: Growing context-sensitive languages and Church-Rosser languages. Inform. Comput. **141**, 1–36 (1998)
5. Hartmanis, J.: Computational complexity of one-tape Turing machine computations. J. ACM **15**, 325–339 (1968)
6. Hennie, F.C.: One-tape, off-line Turing machine computations. Inform. Control **8**, 553–578 (1965)
7. Hibbard, T.N.: A generalization of context-free determinism. Inform. Control **11**, 196–238 (1967)
8. Holzer, M., Jakobi, S., Kutrib, M.: Minimal reversible deterministic finite automata. In: Potapov, I. (ed.) DLT 2015. LNCS, vol. 9168, pp. 276–287. Springer, Heidelberg (2015)
9. Kondacs, A., Watrous, J.: On the power of quantum finite state automata. In: Foundations of Computer Science (FOCS 1997), pp. 66–75. IEEE Computer Society (1997)
10. Kutrib, M.: Aspects of reversibility for classical automata. In: Calude, C.S., Freivalds, R., Kazuo, I. (eds.) Computing with New Resources. LNCS, vol. 8808, pp. 83–98. Springer, Heidelberg (2014)
11. Kutrib, M., Malcher, A.: When Church-Rosser becomes context free. Int. J. Found. Comput. Sci. **18**, 1293–1302 (2007)
12. Kutrib, M., Malcher, A.: Fast reversible language recognition using cellular automata. Inform. Comput. **206**(9–10), 1142–1151 (2008)
13. Kutrib, M., Malcher, A.: Reversible pushdown automata. J. Comput. Syst. Sci. **78**, 1814–1827 (2012)
14. Kutrib, M., Malcher, A.: One-way reversible multi-head finite automata. In: Glück, R., Yokoyama, T. (eds.) RC 2012. LNCS, vol. 7581, pp. 14–28. Springer, Heidelberg (2013)

15. Kutrib, M., Malcher, A., Wendlandt, M.: Reversible queue automata. In: Non-Classical Models of Automata and Applications (NCMA 2014), vol. 304, pp. 163–178. Austrian Computer Society, Vienna (2014). books@ocg.at

16. Kutrib, M., Messerschmidt, H., Otto, F.: On stateless two-pushdown automata and restarting automata. Int. J. Found. Comput. Sci. **21**, 781–798 (2010)

17. Kutrib, M., Wendlandt, M.: On simulation costs of unary limited automata. In: Descriptional Complexity of Formal Systems (DCFS 2015). LNCS. Springer (2015, to appear)

18. Kutrib, M., Worsch, T.: Degrees of reversibility for DFA and DPDA. In: Yamashita, S., Minato, S. (eds.) RC 2014. LNCS, vol. 8507, pp. 40–53. Springer, Heidelberg (2014)

19. Landauer, R.: Irreversibility and heat generation in the computing process. IBM J. Res. Dev. **5**, 183–191 (1961)

20. Lange, K.J., McKenzie, P., Tapp, A.: Reversible space equals deterministic space. J. Comput. System Sci. **60**, 354–367 (2000)

21. McNaughton, R., Narendran, P., Otto, F.: Church-Rosser Thue systems and formal languages. J. ACM **35**, 324–344 (1988)

22. Morita, K.: Reversible computing and cellular automata - a survey. Theoret. Comput. Sci. **395**, 101–131 (2008)

23. Morita, K.: Two-way reversible multi-head finite automata. Fund. Inform. **110**, 241–254 (2011)

24. Morita, K.: A deterministic two-way multi-head finite automaton can be converted into a reversible one with the same number of heads. In: Glück, R., Yokoyama, T. (eds.) RC 2012. LNCS, vol. 7581, pp. 29–43. Springer, Heidelberg (2013)

25. Niemann, G., Otto, F.: The Church-Rosser languages are the deterministic variants of the growing context-sensitive languages. Inform. Comput. **197**, 1–21 (2005)

26. Pighizzini, G., Pisoni, A.: Limited automata and regular languages. Int. J. Found. Comput. Sci. **25**, 897–916 (2014)

27. Pighizzini, G., Pisoni, A.: Limited automata and context-free languages. Fund. Inform. **136**, 157–176 (2015)

28. Pin, J.E.: On reversible automata. In: Simon, I. (ed.) LATIN 1992. LNCS, vol. 583, pp. 401–416. Springer, Heidelberg (1992)

29. Průša, D.: Weight-reducing Hennie machines and their descriptional complexity. In: Dediu, A.-H., Martín-Vide, C., Sierra-Rodríguez, J.-L., Truthe, B. (eds.) LATA 2014. LNCS, vol. 8370, pp. 553–564. Springer, Heidelberg (2014)

An Intrinsically Universal Family of Causal Graph Dynamics

Simon Martiel[(✉)] and Bruno Martin

University of Nice Sophia Antipolis, I3S-CNRS, UMR 7271,
BP121, 06903 Sophia Antipolis, France
simon.martiel@gmail.com, Bruno.Martin@unice.fr

Abstract. Causal Graph Dynamics generalize Cellular Automata, extending them to bounded degree, time varying graphs. The dynamics rewrites the graph in discrete time-steps, with respect to two physics-like symmetries: causality (there exists a bounded speed of information propagation) and shift-invariance (the rewriting acts everywhere the same). Intrinsic universality is the ability of the instance of a model to simulate all other instances, while preserving the structure of the computation. We present here an intrinsically universal family of Causal Graph Dynamics, and give insight on why it seems impossible to improve this result to the existence of a unique intrinsically universal instance.

1 Introduction

Cellular Automata (CA) consist in an array of cells, each of them taking a state in a finite set. The array evolves in discrete time-steps with respect to certain physics-like symmetries: causality (there exists a bounded speed of information propagation) and shift-invariance (the evolution acts everywhere the same). It can be shown that these transformations can be described by a local rule, updating the state of a cell according to the states of its neighbours, applied simultaneously on every cell. Eventhough their origin lie in physics, CA have been studied in many fashions as a model of distributed computation (self-replicating machines, synchronization problems, ...), as well as a variety of multi-agent systems (traffic jam, demographics, ...). Various generalization of this model have been studied: stochastic, asynchronous or non-uniform CA, CA over Cayley graphs, over fixed graphs, Quantum Cellular Automata.

All these generalizations are based on a fixed topology for the configuration's space of the automata. There are many situations, however, in which the notion of 'who is next to whom' also varies in time (e.g. agents become physically connected, get to exchange contact details, move around, etc.). In the literature, several models (of physical systems, self-replication, biochemical agents, economical agents, social networks...) feature such neighbour-to-neighbour interactions with time-varying neighbourhood, thereby generalizing CA to their specific sake. Recently, CA have been generalized to arbitrary, bounded-degree, time-varying graphs which have been studied for their own sake, under the name of *Causal Graph Dynamics* (CGD). See [1,5] for theoretical foundations of CGD.

© Springer International Publishing Switzerland 2015
J. Durand-Lose and B. Nagy (Eds.): MCU 2015, LNCS 9288, pp. 129–148, 2015.
DOI: 10.1007/978-3-319-23111-2_9

Intrinsic universality is the property of having one instance of the model of computation able to simulate all other instances while preserving the structure of the computation. This notion of preserving the structure of the computation has a precise meaning when studying models where a notion of space exists and has already been intensively studied in the cases of CA [6,8,10] or quantum CA [2,3]. In a previous work [7], the authors already proved a preliminary result of intrinsical universality of CGD, namely that, given a description of a graph and a local rule, one can describe an instance of the model constructing the graph and simulating the rule. In this paper, the authors present a family of intrinsically universal local rules, such that, given any local rule on any set of configurations, there exists a rule in this family intrinsically simulating it.

The paper is organized as follows. In Sect. 2, graphs and causal graphs dynamics are introduced, together with some examples. In Sect. 3, a definition of intrinsic universality is proposed. Section 4, offers some preliminary results to facilitate the construction of the family of intrinsically universal rules. Section 5 describes the two encodings used to encode local rule and graphs, and the construction of the family of intrinsically universal rule. In the last section, a discussion of the result and a conclusion are provided.

2   Graphs and Localizable Dynamics

Graphs. Our CGD are over certain kinds of graphs, referred to as *generalized Cayley graphs* which, basically, correspond to the usual, connected, undirected, countable size, bounded-degree graphs, with five added twists:

- Edges are between ports of vertices, rather than between vertices themselves, so that each vertex can distinguish its different neighbours, via the port that connects to it.
- There is a privileged pointed vertex playing the role of an origin, so that any vertex can be referred to relative to the origin, via a sequence of ports leading to it.
- The graphs are considered modulo isomorphism, so that only the relative position of the vertices can matter.
- The vertices and edges are given labels taken in finite sets, so that they may carry an internal state just like the cells of a CA.
- The labelling functions are partial, so that we may express our partial knowledge about part of a graph. For instance is common that a local function may yield a vertex, its internal state, its neighbours, and yet have no opinion about the internal state of those neighbours.

See [5] for a proper formalization of generalized Cayley graphs. Figure 1 shows the differences between graphs, pointed graphs, and generalized Cayley graphs.

Some notations. The *vertices* of the graphs (Fig. 1(a)) we consider in this paper are uniquely identified by a name like u. They may also be labelled with a *state* $\sigma(u)$ in Σ, a finite set. Each vertex has *ports* in a finite set π. A vertex and its port are written $u\!:\!a$. An *edge* is an unordered pair $\{u\!:\!a, v\!:\!b\}$. Such an edge connects

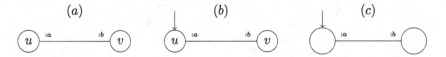

Fig. 1. *The different types of graphs.* (a) A graph G. (b) A pointed graph (G, u). (c) A pointed graph modulo or "Generalized Cayley graph". These are anonymous: vertices have no name and can only be distinguished using the graph structure.

vertices u and v; we shall consider connected graphs only. Because the port of a vertex can only appear in one edge, the degree of the graphs is bounded by $|\pi|$. Edges may also be labelled with a *state* $\delta(\{u\!:\!a, v\!:\!b\})$ in Δ, a finite set. The set of all generalized Cayley graphs (see Fig. 1(c)) of ports π, vertices labels Σ and edge labels Δ is denoted $\mathcal{X}_{\pi, \Sigma, \Delta}$. The set of all classical graphs (see Fig. 1(a)) of ports π, vertices labels Σ and edge labels Δ is denoted $\mathcal{G}_{\pi, \Sigma, \Delta}$.

Paths and Vertices. A Generalized Cayley Graph $X \in \mathcal{X}_{\pi, \Sigma, \Delta}$ is such that any vertex of the graph is identified by the set of paths from the origin to this vertex: for any $u, v \in V(X)$, there is an edge $e = \{u : a, v : b\}$ between u and v if and only if $u.ab \subseteq v$, i.e. any path from the origin to u augmented with the edge e is a path from the origin to v. As a consequence, the origin is the only vertex which contains ε (the empty word). Notice that each vertex can be identified by a particular path from the origin rather than all paths from the origin, for instance by the smallest path according to the lexicographic order. According to this convention the origin would identified by ε. For convenience, from now on, a vertex, i.e. a set of paths, and a path representing this vertex will no longer be distinguished. I.e. we shall speak of "vertex" u in $V(X)$ (or simply $u \in X$).

Operations. For a generalized Cayley graph X (see [5] for details):

- the neighbours of radius r are just those vertices which can be reached with a path of length r starting from the origin,
- the disk of radius r, written X^r, is the subgraph induced by the neighbours of radius $r + 1$, with labellings restricted to the neighbours of radius r and the edges between them.

We denote by X_u the graph whose vertices are named relatively to some other vertex u as the origin. Formally, this is obtained by taking a pointed graph non-modulo the equivalence class X, moving the pointer to u, and then considering the equivalence class again. This graph is referred to as X *shifted by u*.

The composition of a shift, followed by a restriction, applied on X, will simply be written X_u^r. Given a generalized Cayley graph X, and a vertex $u \in X$, we call $\bar{u}$ the inverse path to u. We have $X_{u.\bar{u}} = X$.

Moreover, we need a prefixing operation acting on graphs from the set $\mathcal{G}_{\pi, \Sigma, \Delta}$. In the following definitions, $u.G$ with $u \in \pi^*$ and G a graph, stands for the graph G where names of vertices are prefixed with u.

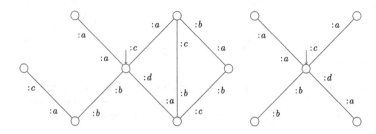

Fig. 2. *A generalized Cayley graph and its disk of radius* 0. *Notice that the set of paths describing vertices in* X^0 *are strict subsets of those in* X, *even though their shortest representative is the same. For instance the path* $ca.cb$ *is in the set whose shortest representative is* da *in* X *but is not a path in* X^0.

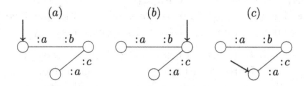

Fig. 3. (a) A generalized Cayley graph X. (b) X_{ab} the generalized Cayley graph X shifted on vertex ab. (c) $X_{ab.ca}$ the generalized Cayley graph X shifted on vertex $ab.ca$, which also corresponds to the graph X_{ab} shifted on vertex ca.

Once given two graphs G and H from the set $\mathcal{G}_{\pi,\Sigma,\Delta}$, it is possible to check if their labelling and ports do not contradict, and to compute their union. If G and H agree on their intersection we say that they are consistent and we denote their union by $G \cup H$. We say that G and H are trivially consistent if their intersection is empty. All these notations are rigorously formalized in [5] (Figs. 2 and 3).

We can now introduce our notion of *local rule*. In a graph generated by a local rule f, names of vertices have a particular meaning. When applied on a disk $X_u^r \in \mathcal{X}_{\pi,\Sigma}^r$, f produces a graph $f(X_u^r)$ such that the names of its vertices are sets of elements of the form $u.z$ with u a path of X_u^r and z a suffix in a set $S = \{\varepsilon, 1, \ldots, b\}$. The conventions taken are such that the integer z stands for the 'successor numbered z'. Hence the vertices designated by $\varepsilon, 1, 2 \ldots$ are successors of the vertex ε, whereas those designated by $u, u.1, u.2 \ldots$ are successors of its neighbour $u \in X^r$. For instance a vertex named $\{1, ab.2\}$ is understood to be both the first successor of vertex ε and the second successor of the vertex attained by the path ab. Such a vertex can be designated by $1, ab.2$ or $\{1, ab.2\}$.

Definition 1 (Local rule). *A (possibly partial) function f from $\mathcal{X}_{\pi,\Sigma}^r$ to $\mathcal{G}_{\pi,\Sigma,\Delta}$ is a local rule if and only if:*

- *For every X, the vertices of $f(X)$ are disjoint subsets of $V(X).S$ and $\varepsilon \in f(X)$,*
- *There exists a bound b such that for all disks X^{r+1}, $|V(f(X^{r+1}))| \leq b$,*

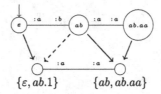

$$\{\varepsilon, ab.1\} \qquad \{ab, ab.aa\}$$

Fig. 4. Naming convention in the image graph of a local rule. The first vertex in the image graph (bottom graph) is both the direct continuation of the first vertex ε and the first successor of the second vertex ab. The second vertex of the image graph is the continuation of both the second vertex ab and the third vertex $ab.aa$. Continuation relation is represented by plain arrows, while successor relation by dashed arrows.

- For every disk X^{r+1} and every $u \in X^0$ we have that $f(X^r)$ and $u.f(X_u^r)$ are non-trivially consistent,
- For every disk X^{3r+2} and every $u \in X^{2r+1}$ we have that $f(X^r)$ and $u.f(X_u^r)$ are consistent (Fig. 4).

The conditions of consistency are here to ensure that if the local rule is applied on two "close" vertices of the same graph, the two resulting subgraphs will be intersecting and consistent.

A local rule is a mathematical object which can be characterized by:

– $|\pi|$ the degree of the graphs it is applied on,
– Σ the set of vertex labels,
– r the radius of the disks it is applied on,
– b the maximal size of its images.

The set of local rules of parameters $(|\pi|, \Sigma, r, b)$ is denoted $\mathcal{F}_{\pi, \Sigma, r, b}$.

Finally, the definition of localizable function describes how these local rules can be used to induce a global function that acts on graphs of arbitrary size.

Definition 2 (Localizable function). *A (global) function F from $\mathcal{X}_{\Sigma, \pi}$ to $\mathcal{X}_{\Sigma, \pi}$ is localizable if and only if there exists a radius r and a local rule f from $\mathcal{X}_{\Sigma, \pi}^r$ to $\mathcal{G}_{\Sigma, \pi}$ such that for all X, $F(X)$ is given by the equivalence class, with ε taken as the pointer vertex, of the graph*

$$\sim \bigcup_{u \in X} u.f(X_u^r).$$

where $\sim G$ constructs the generalized Cayley graph having the same structure as G, with pointer the vertex with name ε.

We now provide two examples of local rules.

The turtle. This transformation is defined over graphs of degree 1. It switches between the two different graphs of degree one. The corresponding local rule is depicted in Fig. 5.

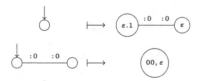

Fig. 5. The turtle local rule. The induced dynamics simply switches between the two existing graphs of ports $\{0\}$. In the first case, two vertices are generated with two "fresh" names. In the second case, a single vertex is generated with name $\{00, \varepsilon\}$. In total two vertices will be generated by the two different disks present in the graph will be identified thanks to the graph union.

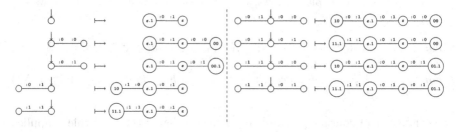

Fig. 6. The inflating line local rule. There are 9 different disks of radius 0 and degree 2. For each of these disks, the local rule generates between 2 and 4 vertices. The vertices named ε and $\varepsilon.1$ are the two direct "descendant" of the center of the disk (the pointed vertex). The other vertices are descendant of the neighbour(s) of the pointed vertex, and are present to allow the recomposition of the image graph through the graph union.

Inflating Line. This transformation is defined over graphs of degree 2, i.e. with ports $\{0, 1\}$. It replaces each vertex by two vertices, doubling the length of the graph. Figure 6 describes the 9 different neighbourhoods of radius 0 and their respective image through a local rule inducing this transformation.

3   Intrinsic Simulation and Universality

When considering the problem of intrinsic simulation inside a model or in-between models, the problem of qualifying the structure of the computation arises. Indeed, intrinsic simulation is about simulating another instance of a model while preserving the structure of the computation. In the case of CA, it is required that one must be able to obtain the simulated configuration by grouping cells of the simulating configuration [11]. In our case, we would like to state that the simulating graph has, somehow, the same topology as the simulated graph. This is done by using the two notions of continuity and shift-invariance.

In [4], it was shown that localizable dynamics introduced in Sect. 2 can also be described in a very axiomatic way by endowing the set of graphs with a compact metric and by defining a notion of shift-invariance for functions over graphs. These definitions can be slightly altered to characterize continuity and shift-invariance for a transformation from a set $\mathcal{X}_{\pi_1, \Sigma_1, \Delta_2}$ to a set $\mathcal{X}_{\pi_2, \Sigma_2, \Delta_2}$.

Definition 3 (Intrinsic simulation). *A localizable dynamics* $(\mathcal{X}_{\pi_1,\Sigma_1}, f_1)$ *intrinsically simulates another localizable dynamics* $(\mathcal{X}_{\pi_2,\Sigma_2}, f_2)$ *if and only if there exists a continuous, shift-invariant, bounded, injective, locally computable function* $E : \mathcal{X}_{\pi_2,\Sigma_2} \to \mathcal{X}_{\pi_1,\Sigma_1}$ *and a constant* δ *such that, for all graph* $X \in \mathcal{X}_{\pi_2,\Sigma_2}$:

$$E \circ F_2(X) = F_1^\delta \circ E(X)$$

Remark on the Notion of Locally Computable Transformations. In [5], it is proved that the application of a continuous shift-invariant transformation, i.e. a CGD, on a finite graph is computable. This property naturally extends to continuous shift-invariant transformations from a set $\mathcal{X}_{\pi_1,\Sigma_1,\Delta_1}$ to a set $\mathcal{X}_{\pi_2,\Sigma_2,\Delta_2}$. This is in fact our notion of locally computable function. Thus, in the previous definition, the condition that E is locally computable is already verified when requiring its continuity and shift-invariance.

Now the definition of intrinsic universality comes naturally:

Definition 4 (Intrinsic universality). *A localizable dynamics* $(\mathcal{X}_{\pi,\Sigma}, f)$ *is intrinsically universal if and only if, it instrinsically simulates any other localizable dynamics* $(\mathcal{X}_{\pi',\Sigma'}, f')$.

Claim 1 (A family of intrinsically universal instances). *There exists a family* $(\mathcal{X}_{\pi_u,\Sigma_u}, f_d)_{d\in\mathbb{N}}$ *such that for all localizable dynamics* $(\mathcal{X}_{\pi,\Sigma,\Delta}, f)$ *there exists an index* d *such that* $(\mathcal{X}_{\pi_u,\Sigma_u}, f_d)$ *intrinsically simulates* $(\mathcal{X}_{\pi,\Sigma,\Delta}, f)$.

Notice that we will not give a proof of this result. The complexity of the construction we present prevents any formal result on its properties. Instead, we will sketch the construction in the following sections.

4   Preliminary Results

Lemmas 1 and 2 are used to limit the set of local rules we need to simulate.

Lemma 1 (Radius 1 is universal). *Let* f *be a local rule of radius* $r = 2^\ell$ *over* $\mathcal{X}_{\pi,\Sigma,\Delta}$. *There exists a local rule* f' *over* $\mathcal{X}_{\pi^r,\Sigma^{\ell+1},\Delta\cup\{\star\}}$ *of radius 1 such that* $(\mathcal{X}_{\pi^r,\Sigma\times\{1,...,\ell\},\Delta\cup\{\star\}}, f')$ *simulates* $(\mathcal{X}_{\pi,\Sigma,\Delta}, f)$.

Proof. Outline. Over the first $i = 1,\dots,\ell$ steps, each vertex will grow some ancillary edges to, in the end, reach all neighbours in its neighbourhood of radius r. More precisely, states of vertices are kept identical, whereas an ancillary edge with state $\star$ is added between any two vertices at distance 2. Moreover, the vertices count until stage ℓ. At this point, the neighbours that were initially at distance r have become visible at distance one. The local rule f can be applied, all ancillary edges are dropped, and all counters are reset.

Lemma 2 (Label free is universal). *Let* f *be a local rule of radius* r *over* $\mathcal{X}_{\pi,\Sigma,\Delta}$. *There exists a local rule* f' *over* $\mathcal{X}_{\pi\cup|\Sigma|\times|\Delta|,\varnothing,\varnothing}$ *such that* $(\mathcal{X}_{\pi',\varnothing,\varnothing}, f')$ *simulates* $(\mathcal{X}_{\pi,\Sigma,\Delta}, f)$, *where* $\pi' = \pi \sqcup \Sigma \sqcup \Delta^{|\pi|}$.

Proof. Outline. The presence of a label $i \in \Sigma$ on a vertex will be encoded by the presence of a dangling vertex on port i of this vertex. In the same fashion, if an edge labelled with $j \in \Delta$ connects two ports $u : a$ and $v : b$, then vertex u will have a dangling vertex on port j in its ath port component and v a dangling vertex in its bth port component. Notice that not all graphs are valid encoding, e.g. if a vertex has a dangling vertex on port $i \in \Sigma$ and on port $j \in \Sigma$ at the same time. Nevertheless this encoding verify all the require properties as it is injective, continuous and shift-invariant.

Notice that these two constructions are not incompatible. Composing the two in the right order leads to the fact that any local rule can be intrinsically simulated by a local rule of radius one with no labels. In other words, the subset of localizable dynamics of radius one with no labels is intrinsically universal.

Corollary 1 (Radius 1 label free is universal). *Let f be a local rule of radius r over $\mathcal{X}_{\pi,\Sigma,\Delta}$. There exists a local rule f' over $\mathcal{X}_{\pi'\varnothing,\varnothing}$ of radius 1 such that $(\mathcal{X}_{\pi',\varnothing,\varnothing}, f')$ simulates $(\mathcal{X}_{\pi,\Sigma,\Delta}, f)$.*

5   A Family of Intrinsically Universal Local Rules

We now describe a family of intrinsically universal local rules $(f_d, \mathcal{X}_{\pi_u,\Sigma_u,\Delta_u})_d$ such that f_d simulates all rules over $\mathcal{X}_{\pi,\varnothing,\varnothing}$ with $|\pi| = d$. More precisely, all of these universal rules will act upon the same set of graphs $\mathcal{X}_{\pi_u,\Sigma_u,\Delta_u}$, and only differ in their radius. To define these rules, we are faced with several problems:

- Our universal rules all act upon a given set of graphs of bounded degree $|\pi_u|$. We need to be able to encode any graphs of bounded degree into our set of graphs $\mathcal{X}_{\pi_u,\Sigma_u,\Delta_u}$. Section 5.1 tackles this issue and introduces an encoding of any graph of bounded degree in a graph of degree 3.
- There is an unbounded number of local rules of radius 1 with no labels. Hence, the information of which local rule is to be simulated cannot be stored as a label in Σ_u. Section 5.2 offers an encoding of any local rule in a subgraph whose purpose is to be attached to every simulated vertex.
- In order to simulate more than a single time step of the local rule, we must be able to create several instances of the graph containing its encoding and transmit these instances to the descendants of the simulated vertex. Section 5.3 offers a way to duplicate a subgraph describing a local rule, together with some synchronization tools.

A description of the functioning of the universal local rule is given in Subsect. 5.3. In this section we might refer to simulated vertices as "meta"-vertices since each of these vertices will be encoded in a graph structure.

5.1   Graph Encoding

We choose the following encoding to represent any graph of bounded degree π in a graph of degree 3. For readability, we will give explicit names to the three

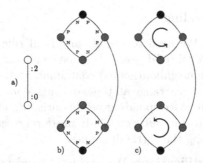

Fig. 7. Here, $\pi = \{0, 1, 2\}$. (a) depicts a graph composed of two vertices connected through ports 2 and 0. (b) represents the encoding of this graph. Each vertex is represented by 4 vertices forming a ring. The darkest vertices have label VERTEX while the grey vertices have label PORT. The ports used in the ring are (N)ext and (P)revious. The edge linking the two vertices is represented by an edge between the third vertex of the first ring (representing port 2 of the first vertex) and the first vertex of the second ring (representing port 0 of the second vertex). Finally, (c) presents a lighter representation of the same encoding where an arrow indicates the orientation of the rings. For the sake of clarity, this latter representation will be used in the following figures.

ports used in the following definition. The set of ports in the encoding can be assimilated to $\{0, 1, 2\}$. The three port are: *previous*, *next* and *neighbour*. The set containing those three ports will be referred to as π_{graph}. We define the set of labels Σ_{graph} as the set $\{VERTEX, PORT\}$.

Definition 5 (Graph encoding). *Given a set of ports π, consider the transformation $E_\pi^{graph} : X_\pi \to X_{\pi_{graph}, \Sigma_{graph}}$ defined as follows:*

- *To each vertex v in X, corresponds $\pi + 1$ vertices $v_0, \ldots, v_\pi$ in $E_\pi^{graph}(X)$ and the following edges: for all $i \in \{0, \ldots, |\pi|\}$, $\{v_i : next, v_{i+1} : previous\}$. v_i has label PORT for $i < |\pi|$ and $v_{|\pi|}$ has label VERTEX.*
- *To each edge $\{u : i, v : j\}$ in X corresponds an edge $\{u_i : neighbour, v_j : neighbour\}$.*

The idea is to split the encoded vertex into $|\pi| + 1$ vertices and arrange them into a ring. Each vertex v_i ($i < |\pi|$) represents a port of the encoded vertex. The last vertex v_π marks the start of the ring (the vertex representing port 0 will be found on its port *next*). Figure 7 describes the encoding for graphs with $|\pi| = 3$.

Lemma 3 (E_π^{graph} is a good encoding). *Given π, E_π^{graph} is continuous, shift-invariant and injective.*

Proof. The proof of this result is pretty straightforward. As E_π^{graph} acts locally on the graph, continuity and shift-invariance are instantaneous. Moreover, any change in the original graph will result in a change in the encoded graph as all information on the topology is preserved.

5.2 Local Rule Encoding

[**General structure**]. We need to encode any local rule of radius 1 without label into a subgraph. A rule of degree $|\pi|$ can be seen as an array of fixed length (the number of possible neighbourhoods) containing all the possible outputs of the local rule. We choose to arrange all these outputs along a line graph together with a description of the corresponding neighbourhood. The description of the neighbourhoods is detailed in Subsect. 5.3. Figure 8 represents such an encoding for the local rule inducing the turtle dynamics.

[**Addresses and Identification**]. We also need to identify a meta-vertex of an output to another meta-vertex in another output, in order to proceed to a graph union. This is done by adding to each vertex labelled by *VERTEX* a line graph containing a path towards the other vertex. Figure 9 represents the graph encoding the turtle local rule with these addresses. Figure 10 represents the graph encoding the inflating line local rule.

[**Inheritors and Disowned Vertices**]. Inside an output subgraph, there are two types of meta-vertices: the ones that need to receive a copy of the local rule graph and the others. We use a product label to mark the meta-vertices that will inherit of a copy of the local rule. In the example of the turtle, all meta-vertices are marked while in the example of the inflating line, only the meta-vertices having an empty address are marked.

5.3 Description of an Intrinsically Universal Rule

Applying a local rule to every vertex in a graph consists in several stages:

(i) Each vertex observes its neighbourhood
(ii) Each vertex deduces the output subgraph it has to produce according to the local rule
(iii) A graph union of all these subgraphs is computed to produce the final graph.

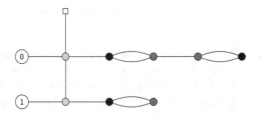

Fig. 8. Encoding of the turtle rule. Black vertices are vertices labelled by VERTEX, dark grey vertices are labelled by PORT. Light grey vertices are part of line structure onto which all outputs are attached. Vertices on the left of the vertical line are labelled by bits corresponding to the number of the outputs in an enumeration of all possible neighbourhoods. We choose not to use the neighbourhood encoding used in Sect. 5.3, as there are only 2 different neighbourhoods. The square vertex represents the top of the line structure.

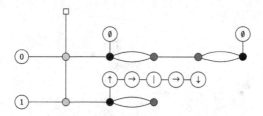

Fig. 9. Encoding of the turtle rule including the addresses. The empty set label is used to specify that the meta-vertex does not need to be identified to another meta-vertex. When the address is not empty, it is encoded in a line graph using 4 different labels: ↑, →, ↓ and |. ↑ indicates to move on the father meta-vertex. ↓ indicates to go down from a father meta-vertex to its output. → indicates to travel along the port NEXT in a meta-vertex. | indicates to travel along the port neighbour between two meta-vertices.

The universal local rule implements those three stages, with an additional stage:

(ii*) The encoding of the local rule is duplicated into each meta vertex of the chosen output subgraph.

Moreover, a universal local rule must synchronise the simulation in every meta-vertex in order to perform the graph union only when all subgraphs are chosen and all duplications are over.

We detail how each of these stages are performed by the universal local rule.

[neighbourhood Observation] First the meta-vertex proceeds to generate a matrix of vertices of size $|\pi| + 1$ to store the connectivity of its neighbourhood. A new vertex is attached to the vertex labelled *VERTEX* and starts moving along the ring of vertices labelled *PORT* growing the matrix in 2 passes. Figure 11 describes this growing process on a meta-vertex of degree 9.

Once the matrix is built, the machine vertex starts a depth first search (DFS) of depth 1 on the meta vertex it is attached on. It grows 2 edges (or arms) that will travel in the graph and 4 unary counters to keep track of the DFS status. The unary counters are line graphs of lengths $|\pi| + 1$ and $|\pi|$. The two counters of length $|\pi| + 1$ keep track of which meta-vertex can be found at the end of each arm while the two counters of length π keep track of the ports currently considered. Figure 13 represents the structure of a counter, and Fig. 14 describes the structure used to store the current state of the DFS. While visiting a vertex u_i, its port p_j and a vertex u_k and its port p_l, an edge is created between cells (i, j) and (k, l) of the matrix if the edge $\{u_i : p_j, u_k : p_l\}$ is present in the graph. Once the DFS is over, the matrix contains enough information to determine the neighbourhood of the vertex. Figure 12 presents the two different matrices for neighbourhoods in graphs of degree 1.

Note on the Description of the Neighbourhood. The usage of a matrix to encode the neighbourhood of a meta-vertex is the most general solution we can implement. However, in most of the cases, we do not need that much information. In the two examples we develop in this paper (the turtle and the inflating line), it is only required to test for the existence of a potential neighbour on each port

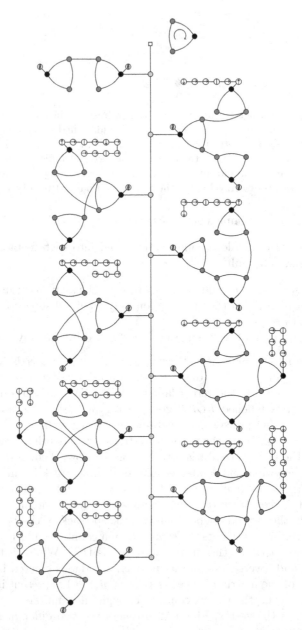

Fig. 10. Encoding of the inflating line rule including the addresses. There exist 9 different neighbourhoods of radius 0 on graphs of degree 2, thus the presence of 9 different outputs in the encoding. Numbering of the possible outputs are neglected here as they do not bring any information to the understanding of this encoding.

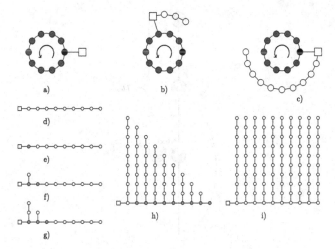

Fig. 11. Growth of the connectivity matrix. A "machine" vertex starts to run along the ring and for each vertex it passes, adds a new vertex to a line graph: (a) At first the line graph is empty and the machine vertex is attached on the VERTEX vertex. (b) After three steps. (c) After 10 steps, the machine vertex is back on the first vertex and start the second pass. d,e,f,g represent the 4 first steps of the second pass. The machine sends a signal (in grey) that triggers the growth of each column while moving along the ring. (h) represents the 11th step where the signal reaches the last column and the machine arrives at the VERTEX label again. The machine sends an "end" signal to stop the growth of the columns. (i) represents the final matrix (after 20 steps).

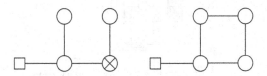

Fig. 12. The two matrices encoding resp. the neighbourhood where no neighbour is present, and the neighbourhood where another vertex is present on the only port. In the first graph, the bottom right vertex is crossed to indicate that there is no "second" vertex in the neighbourhood.

of the meta-vertex, and the local rule does not require to know their complete connectivity. Hence, in both local rule encodings, we will use an ad hoc encoding of the neighbourhoods. For the turtle rule, we will use a single bit to represent the two possible neighbourhoods. For the inflating line, a string of bits is used. For each port of the vertex we proceed as follows: if there is no neighbour on this port we add a 0 to the string. If there is a neighbour on the port, we add a 1 to the string, followed by a 0 or a 1 depending on the port and the other end of the edge.

[**Choosing the Output Subgraph**]. After recording the local connectivity, the machine vertex starts to travel down the local rule encoding and compare this

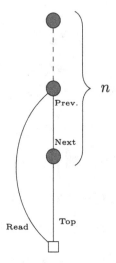

Fig. 13. A counter structure. It consists in a line graph of the appropriate length. The origin of the line can grow an arm to read the counter, one vertex at a time. It is easy for an automaton to grow a counter of the appropriate size by running along a meta-vertex and generating a new vertex for each visited port (see matrix generation). All vertices composing the counter have the same label.

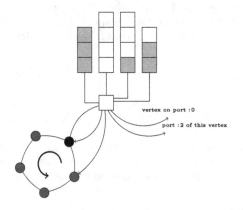

Fig. 14. DFS structure for the neighbourhood exploration of a vertex of degree $|\pi| = 4$. At the top: 4 unary counters structure. The DFS explores the neighbourhood of the center vertex by considering every possible pair of vertices (including the center vertex, as there might be loops the graph). The two center counters are used to keep track of which vertices are currently being visited. The two smaller counters are here to keep track of which port is currently considered in each of these vertices. Here, the currently considered pair is composed of the "center" vertex and its neighbour of port :0 and their ports :3 and :2.

recording to the information attached to each output, stopping when the two are matching.

[**Duplicating the Local Rule Encoding**]. The machine then initiates a DFS on the chosen output graph. The purpose of this DFS is to search for marked meta-vertices. During the DFS, every time a marked meta-vertex is met, the DFS is paused and a new DFS starts from the root of the local rule encoding. This new DFS will explore the local rule encoding while constructing a new copy of it. This can be done by maintaining a stack structure containing the path followed in the graph during the DFS. When the machine encounters an edge leading toward a previously visited vertex, it uses this stack to backtrack and find the right edge to create in the new version of the graph. These two DFS act on graphs of degree 3 and 4 and thus do not require the same counter structures as the DFS in the neighbourhood observation stage, as everything can be stored using a bounded number of labels in the vertices. This new version of the local rule encoding is then attached to the marked meta-vertex and the first DFS is resumed.

[**Graph Union and Vertex Identification**]. Once the DFS is over, meta-vertices of the output graph start moving in the graph according to the addresses attached to them. Meanwhile, the local rule encoding is reduced to get rid of all the unused outputs, leaving the only chosen output attached to the simulated meta-vertex.

[**Merging Two Meta-Vertices**]. After moving according to its address, a meta-vertex will meet its target meta-vertex and they will try to merge. Notice that the target meta-vertex might also try to merge with a third vertex, and so on, forming a sequence of meta-vertices that must be merged in a single meta-vertex. This can lead to two very distinct situation: either the sequence is not cyclic or the sequence forms a cycle. In the first case, the first vertex of the sequence will perform the merging, followed by the second and so on until all the meta-vertices are merged as a single meta-vertex. In the second case, no meta-vertex can decide to start the merging process as every meta-vertex sees itself in the middle of the sequence. Meta-vertices can easily decide whether this is the case by growing a new edge whose extremity will travel along the sequence. If the edge reaches the end of the sequence, then the merging process will start. If not, a synchronization process will start.

[**Synchronization Process**]. If synchronization is required during the merging process, then we can assume that these meta-vertices are synchronized (i.e. they decide to start the merging process exactly at the same time). Indeed, if they were not synchronized, then the symmetry could have been broken in the local rule encoding, as only the neighbourhood of simulated meta-vertex has an influence on the time step at which meta-vertices of its local rule encoding decide to merge. Two problems now arise:

- In order for the cycle to collapse in a single meta-vertex in a single time step, the universal local rule must be able to "see" the whole cycle, hence its radius must be of at least half the length of the larger possible identification cycle.
- Meta-vertices are not composed of a single vertex. They contain at least $|\pi|+1$ vertices and might also contain a local rule encoding. All these vertices need

to be simultaneously merged with their corresponding vertices in the previous and next meta-vertices in the merging cycle.

The first problem is easy to solve as we are constructing a family of intrinsically universal local rules. A given local rule can only produce merging cycles of bounded lengths, hence will be simulated by one of our universal local rule.

The second problem can be solved using a solution of a problem known as the Firing Squad Synchronisation Problem (FSSP) over graph automata [9,12]. The construction uses labels on the vertices of a graph in order to synchronize all the vertices using only local communications between vertices. Moreover, the solution only depends on the degree of the graph to synchronize. In our case, we need to synchronize meta-vertices and their local rule encodings, which are of bounded degree 4. The identification process will be performed as follow:

- Meta-vertices will detect that they are in a cycle of identification
- Meta-vertices start a FSSP on their main vertex
- The FSSP synchronizes every vertex composing the meta-vertex and its potential local rule encoding
- While propagating the FSSP, new edges are built between vertices of the meta-vertex and their corresponding vertices in the previous and next meta-vertices
- When all vertices are synchronized, the universal local rule performs a merging of the vertices, leading to a single meta-vertex and its local rule encoding.

Figures 16 and 15 describe the different possible cases of merging sequences and the synchronization process. When all mergings are performed, the original meta-vertices are destroyed, leaving only the new graph and can be restarted to simulate the next time-step. Figures 17 and 18 describe the complete simulation of one time step of the turtle dynamics over the graph containing two vertices.

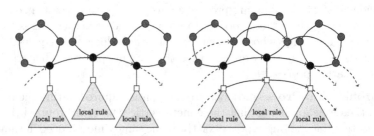

Fig. 15. Synchronization process of three meta-vertices and their local rule encodings. All meta-vertices start a FSSP on their vertices. At the beginning, only the "main" vertex is connected along the merging cycle to the others "main" vertices (top graph). As the FSSP is propagated, the vertices connect themselves to their corresponding vertices in the previous and next vertices. The bottom graph describes the same graph, two propagation steps later. When the FSSP is completed, all vertices "fire" exactly at the same time and perform a merging along all the built cycles, resulting in a single meta-vertex and its local rule encoding.

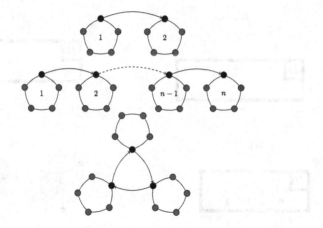

Fig. 16. Different types of merging sequences. The two top cases are solved by ordering the vertices according to the sequence, and then having the first one merge to the second one, and so on. In the last case, the sequence forms a cycle, and a synchronization is necessary to perform the simultaneous merging of the cycle.

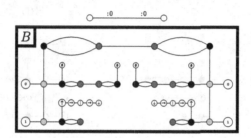

Fig. 17. A graph of degree $|\pi| = 1$ and its encoding together with turtle local rule encodings. Each of the two meta-vertices receive a version the local rule encoding. Once again, the description of the different neighbourhoods in the local rule encoding is not made using matrices, as there are only two possible cases, instead we used single digits. Here the neighbourhood with only one single vertex is encoded by a 0 while the neighbourhood with two vertices is encoded by a 1.

5.4   On the (non) Existence of a Single Universal Rule

The construction presented in this paper describes a family of intrinsically universal local rules, and not an universal local rule. All local rules in this family act on the same set of graphs, and only differ in their radius. Having universal local rules with arbitrary large radius is only required in the last part of the construction, for the merging process. When meta-vertices decide to merge into a single meta-vertex, and the merging sequence forms a cycle, the local rule must be able to either order the meta-vertices and proceed to merge them one-by-one according to that order, or "see" all the meta-vertices, synchronize them, and proceed to the merging in one time step. The latter is only possible if the radius of the local rule is large enough, and that is the solution we adopted here. In the

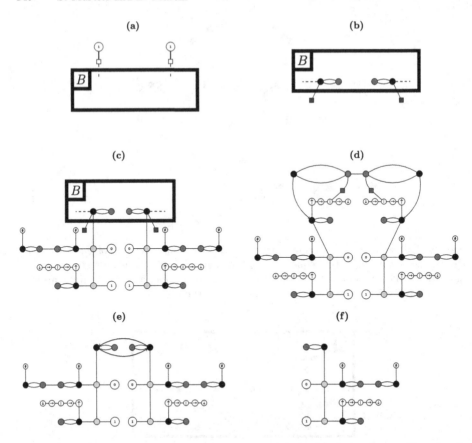

Fig. 18. Steps of the simulation of the turtle local rule. (a) After the neighbourhood exploration. The two automata attached to each meta-vertices have detected the presence of another meta-vertex in the neighbourhood, and thus generated a vertex labelled 1. (b) The automata travelled down the local rule and chose the second output subgraph. They will then start a DFS on the chosen subgraph. (c) the DFS detected a meta-vertex and decided to duplicate the local rule and attach a copy to it. (d) The DFS is over, the local rule is destroyed and the identification process is running. The automata is on his way to reach the meta-vertex at the end of the address attached to the meta-vertex of the output subgrah. 2 symbols of the address have been read: ↑ and →. (e) After reading the address the two meta-vertices are pointing toward each other and start the merging process. (f) After synchronization, the two meta-vertices and their local rule encodings are merged, and the simulation is over. To restart the simulation a new automaton can be attached to the meta-vertex.

former however, we must order meta-vertices that are descendant of different vertices of the simulated graph. This requires to be able to unambiguously order the meta-vertices in any disk of radius 1 of our simulated graph. This is equivalent to have a clean colouring of the simulated graph. The colouring will then give us a way to totally order the descendants of the meta-vertices, and hence gives us a way to proceed to the merging without requiring any synchronization.

However, to use a clean colouring, we must prove that any local rule can be modified to take the colouring into account and maintain it over time. This, in turn, requires to be able to locally break the symmetries in the image graph, which might be impossible for some graphs.

Hence, it seems impossible to construct a unique intrinsically universal rule, at least using this type of constructions.

6   Conclusion

In this work, we provide a definition of intrinsical simulation and intrinsical universality for causal graph dynamics. We then construct a family of intrinsically universal instances of this model. All the local rule of this family act on the same set of graphs and only differ in their radius.

This construction is not optimal, and can still be optimized in various ways:

- One could achieve a similar result with a construction on graphs of smaller degree, and with a smaller label set
- The time-complexity of the simulation can probably be decreased by optimizing the structure of the graph encoding and the local rule encoding. For instance one could imagine using a set structure to encode the local rule, changing a linear access time (in the number of possible neighbourhoods) to the chosen output graphs in a logarithmic access time.

Moreover, it seems that this is the best result we can achieve with this kind of construction, as it seems impossible to construct a unique intrinsically universal rule. Notice that, although we need a family of instances to simulate all the possible instances, most of the "natural" instances can be simulated by the universal rule that allows merging sequence of arbitrary length and only forbids merging cycles of length greater that 4, i.e. the universal local rule of radius 2.

References

1. Arrighi, P., Dowek, G.: Causal graph dynamics. In: Czumaj, A., Mehlhorn, K., Pitts, A., Wattenhofer, R. (eds.) ICALP 2012, Part II. LNCS, vol. 7392, pp. 54–66. Springer, Heidelberg (2012). doi:10.1007/978-3-642-31585-5_9
2. Arrighi, P., Fargetton, R., Wang, Z.: Intrinsically universal one-dimensional quantum cellular automata in two flavours. Fundam. Inform. **21**, 1001–1035 (2009). doi:10.3233/FI-2009-0041
3. Arrighi, P., Grattage, J.: Partitioned quantum cellular automata are intrinsically universal. Nat. Comput. **11**, 13–22 (2012). doi:10.1007/s11047-011-9277-6
4. Arrighi, P., Martiel, S.: Generalized Cayley graphs and cellular automata over them. In: Proceedings of GCM 2012, pp. 129–143, Bremen, September 2012. arXiv:1212.0027
5. Arrighi, P. Martiel, S. Nesme, V.: Generalized Cayley graphs and cellular automata over them. submitted (long version) (2013). arXiv:1212.0027

6. Durand-Lose, J.O.: Intrinsic universality of a 1-dimensional reversible cellular automaton. In: Reischuk, R., Morvan, M. (eds.) STACS 1997. LNCS, vol. 1200, pp. 439–450. Springer, Heidelberg (1997). doi:10.1007/BFb0023479

7. Martiel, S., Martin, B.: Intrinsic universality of causal graph dynamics. In: Neary, T., Cook, M. (eds.) Proceedings, Machines, Computations and Universality 2013, pp. 137–149, Zürich, Switzerland, 09 September 2013–11 September 2013, Electronic Proceedings in Theoretical Computer Science 128. Open Publishing Association (2013). doi:10.4204/EPTCS.128.19

8. Martin, B.: Cellular automata universality revisited. In: Chlebus, B.S., Czaja, L. (eds.) FCT 1997. LNCS, vol. 1279, pp. 329–339. Springer, Heidelberg (1997)

9. Mazoyer, J.: An overview of the firing squad synchronization problem. In: Choffrut, C. (ed.) LITP 1986. LNCS, vol. 316, pp. 82–94. Springer, Heidelberg (1988)

10. Ollinger, N.: Intrinsically universal cellular automata. In: Neary, T., Woods, D., Seda, A.K., Murphy, N. (eds.) CSP, pp. 259–266. Cork University Press (2008). doi:10.4204/EPTCS.1.19

11. Ollinger, N.: Intrinsically universal cellular automata. In: Neary, T., Woods, D., Seda, A.K., Murphy, N. (eds.) CSP, EPTCS 1, pp. 199–204. http://arxiv.org/abs/0906.3213

12. Rosenstiehl, P., Fiksel, J.R., Holliger, A., et al.: Intelligent graphs: networks of finite automata capable of solving graph problems. In: Read, R.C. (Ed.) Graph Theory and Computing, pp. 219–265. Academic Press, Edinburg (1972)

The Simulation Powers and Limitations of Hierarchical Self-Assembly Systems

Jacob Hendricks[✉], Matthew J. Patitz, and Trent A. Rogers

Deptartment of Computer Science and Computer Engineering,
University of Arkansas, Fayetteville, AR, USA
{jhendric,patitz,tar003}@uark.edu

Abstract. In this paper, we extend existing results about simulation and intrinsic universality in a model of tile-based self-assembly. Namely, we work within the 2-Handed Assembly Model (2HAM), which is a model of self-assembly in which assemblies are formed by square tiles that are allowed to combine, using glues along their edges, individually or as pairs of arbitrarily large assemblies in a hierarchical manner, and we explore the abilities of these systems to simulate each other when the simulating systems have a higher "temperature" parameter, which is a system wide threshold dictating how many glue bonds must be formed between two assemblies to allow them to combine. It has previously been shown that systems with lower temperatures cannot simulate arbitrary systems with higher temperatures, and also that systems at some higher temperatures can simulate those at particular lower temperatures, creating an infinite set of infinite hierarchies of 2HAM systems with strictly increasing simulation power within each hierarchy. These previous results relied on two different definitions of simulation, one (*strong* simulation) seemingly more restrictive than the other (*standard* simulation), but which have previously not been proven to be distinct. Here we prove distinctions between them by first fully characterizing the set of pairs of temperatures such that the high temperature systems are intrinsically universal for the lower temperature systems (i.e. one tile set at the higher temperature can simulate any at the lower) using strong simulation. This includes the first impossibility result for simulation downward in temperature. We then show that lower temperature systems which cannot be simulated by higher temperature systems using the strong definition, can in fact be simulated using the standard definition, proving the distinction between the types of simulation.

1 Introduction

In computational theory, a powerful and widely used tool for determining the relative powers of systems is simulation. For instance, in order to prove the

J. Hendricks, M.J. Patitz, and T.A. Rogers—Supported in part by National Science Foundation Grant CCF-1422152.

T.A. Rogers—This author's research was supported by the National Science Foundation Graduate Research Fellowship Program under Grant No. DGE-1450079.

© Springer International Publishing Switzerland 2015

J. Durand-Lose and B. Nagy (Eds.): MCU 2015, LNCS 9288, pp. 149–163, 2015.
DOI: 10.1007/978-3-319-23111-2_10

equivalence (in terms of computational power) of Turing machines and various abstract models (such as tag systems, counter machines, cellular automata, and tile-based self-assembly models), systems have been developed in each which demonstrate their abilities to simulate arbitrary Turing machines, and vice versa. This has been used to prove that whatever can be computed by a system within one model can also be computed by a system in another. Additionally, the notion of a universal Turing machine is based upon the fact that there exist Turing machines which can simulate others.

The methods of simulation which are typically employed involve mappings of behaviors and states in one model or system to those in another, often following some "natural" mapping function, and also often in such a way that the simulation is guaranteed to generate the same final result as the simulated system, and maybe even some or all of its intermediate states. Nonetheless, there is usually no requirement that the simulator "do it the same way," i.e. the dynamical behavior of the simulator need not mirror that of the simulated. For instance, as one Turing machine A simulates another, B, its head movements may be in a significantly different pattern than B's since, for instance, it may frequently move to a special portion of the tape which encodes B's transition table, then back to the "data" section.

While such types of simulation can be informative when asking questions about the equivalence of computational powers of systems, oftentimes it is the behavior of a system which is of interest, not just its "output." Self-assembling systems, which are those composed of large numbers of relatively simple components which autonomously combine to form structures using only local interactions, often fall into this category since the actual ways in which they evolve and build structures are of key importance. In this paper, we focus our attention on tile-based self-assembling systems in a model known as the 2-Handed Assembly Model (2HAM) [3], which is a generalization of the abstract Tile Assembly Model (aTAM) [19] in which the basic components are square *tiles* which are able to bind to each other when they possess matching *glues* on their edges. In the aTAM, assembly occurs as tiles autonomously combine, with one tile at a time attaching to a growing assembly. In the 2HAM, similar growth can occur, but it is possible for pairs of arbitrarily large assemblies (a.k.a. supertiles) to combine as well. Because the dynamical behaviors of these systems are of such importance, work in these models (e.g. [8,9,12,13,16,20]) has turned to a notion of simulation developed within the domain of cellular automata, whose dynamical behaviors are also often of central importance. This notion of simulation, called *intrinsic universality* (see [1,2,5,6,10,11,15,17,18] for some examples related to various models such as cellular automata), is defined in such a way that the simulations performed are essentially "in place" simulations which mirror the dynamics of the simulated systems, modulo a scale factor allowed the simulator. Intrinsic universality has been used to show the existence of "universal" systems, somewhat analogous to universal Turing machines, which can simulate all other systems within a given model or class of systems, but in a dynamics-preserving way. Previous work [9] has shown that there exists a single aTAM tile set which is capable of simulating any arbitrary aTAM system, and thus that tile set is intrinsically universal (IU) for the aTAM (and we also say that the aTAM is IU). Further work in [8] showed that the 2HAM is much more complicated in terms of IU, with there existing hierarchies of 2HAM systems with strictly-increasing power of simulation. These simulations are performed by scaled blocks of tiles known as macrotiles in the simulator used to simulate individual tiles in the simulated systems. The simulation hierarchy in the 2HAM is based on a classification of systems separated by a system parameter known as the *temperature*, which is the global threshold that specifies the minimum strength of glue bindings required for pairs of tiles or supertiles to combine. It was proven in [8] that for every temperature $\tau \geq 2$, there exists a system at temperature τ such that no system at temperature $\tau' < \tau$ can simulate it. However, they also showed that for each $\tau \geq 2$, the class of 2HAM systems at τ is IU.

The motivation of the current paper is to extend and further develop the results of [8], especially Theorem 4 which states: "There exists an infinite number of infinite hierarchies of 2HAM systems with strictly-increasing power (and temperature) that can simulate downward within their own hierarchy." Our results elucidate more details about this hierarchy, including proving important differences between different notions of simulation used to characterize intrinsically universal systems. More specifically, different definitions of simulation have been used even within the IU results of [8], with one referred to as *strong* simulation and one as (*standard*) simulation. Strong simulation is a stricter notion essentially stating that whenever two supertiles in the simulated system $\mathcal{T}$ can combine, every pair of macrotiles that represents them in the simulator $\mathcal{S}$ must be able to (eventually) combine. However, standard simulation simply requires that for each half of such a pair in the simulator, there must exist some mate with which it can eventually combine. While both notions of simulation were utilized in [8], no concrete distinction was proven in terms of what is or isn't possible between them. Here, we first prove that higher temperature systems can strongly simulate lower temperature systems if and only if there is a relationship between the temperature values which we call a *uniform mapping*. We show that it is easy to find whether such a mapping exists between two temperatures and, if so, what one is, and prove that for each pair of temperatures $2 < \tau < \tau'$ where a uniform mapping exists from τ to τ', that there exists a tile set which, at temperature τ', is IU for the class of 2HAM systems at τ. We then prove that if no uniform mapping exists from τ to τ', then there exist systems at τ which cannot be strongly simulated by any system at τ', which is the first impossibility result for simulating downward in temperature that we are aware of, and is of interest because a natural intuition is that higher temperature systems are strictly more powerful. (However, we also show that for any given τ there are only a finite number of $\tau' > \tau$ to which a uniform mapping does not exist.) Finally, we show that some systems which cannot be strongly simulated by higher temperature systems when no uniform mapping exists between temperatures can in fact be simulated from the higher temperature using the standard definition of simulation. This shows the first clear distinction between what is possible under the various definitions, and that the notion of strong simulation is provably more restrictive than that of (standard) simulation since the set of systems which can be simulated by a higher temperature system is strictly greater than that which can be strongly simulated.

In the next section we provide the definitions of the model and framework for our results, then provide an overview of our results in the following sections. Please note that due to space constraints, proofs can be found in an extend version of the paper [14].

2   Definitions

2.1   Informal Definition of the 2HAM

Here we give a brief, informal, sketch of the 2HAM. Please see [14] for a more formal definition. The 2HAM [4,7] is a generalization of the aTAM [19], and in both the basic components are "tiles". A *tile type* is a unit square with four sides, each having a *glue* consisting of a *label* (a finite string) and *strength* (a non-negative integer). We assume a finite set T of tile types, but an infinite number of copies of each tile type, each copy referred to as a *tile*. A *supertile* is (the set of all translations of) a positioning of tiles on the integer lattice $\mathbb{Z}^2$. Two adjacent tiles in a supertile *interact* if the glues on their abutting sides are equal and have positive strength. Each supertile induces a *binding graph*, a grid graph whose vertices are tiles, with an edge between two tiles if they interact. The supertile is τ-*stable* if every cut of its binding graph has strength at least τ, where the weight of an edge is the strength of the glue it represents. That is, the supertile is stable if at least energy τ is required to separate the supertile into two parts. A 2HAM *tile assembly system* (TAS) is a triple $\mathcal{T} = (T, S, \tau)$, where T is a finite tile set, S is a set of *seed* supertiles over T, and τ is the *temperature*, usually 1 or 2. When S is solely an infinite number of each of the singleton tiles of T, we call that the default initial state, and for shorthand notion refer to a TAS with a default initial state simply as a pair $\mathcal{T} = (T, \tau)$. Given a TAS $\mathcal{T} = (T, S, \tau)$, a supertile is *producible*, written as $\alpha \in \mathcal{A}[\mathcal{T}]$ if either it is a (super)tile in S, or it is the τ-stable result of translating two producible assemblies without overlap. That is, any τ-stable supertile which can result from some positioning of two producible supertiles, so that they do not overlap and they bind with at least strength τ, is itself a producible supertile. This potentially allows for the combination of pairs of arbitrary large supertiles. A supertile α is *terminal*, written as $\alpha \in \mathcal{A}_\square[\mathcal{T}]$ if for every producible supertile β, α and β cannot be τ-stably attached.

2.2   Definitions for Simulation

In this subsection, we formally define what it means for one 2HAM TAS to "simulate" another 2HAM TAS. The definitions presented in this (and the next) subsection are based on the simulation definitions from [3,9,16] and are included here for the sake of completeness. We will be describing how the assembly process followed by a system T is simulated by a system U, which we will call the *simulator*. The simulation performed by U will be such that the assembly process followed by U mirrors that of the simulated system T, but with the individual tiles of T represented by (potentially large) square blocks of tiles in U called *macrotiles*. We now provide the definitions necessary to define U as a valid simulator of T. For a tileset T, let A^T and $\tilde{A}^T$ denote the set of all assemblies over T and all supertiles over T respectively. Let $A^T_{<\infty}$ and $\tilde{A}^T_{<\infty}$ denote the set of all finite assemblies over T and all finite supertiles over T respectively.

In what follows, let U be a tile set. An *m-block assembly*, or *macrotile*, over tile set U is a partial function $\gamma : \mathbb{Z}_m \times \mathbb{Z}_m \dashrightarrow U$, where $\mathbb{Z}_m = \{0, 1, \ldots m-1\}$. Let B^U_m be the set of all m-block assemblies over U. The m-block with no domain is said to be *empty*. For an arbitrary assembly $\alpha \in A^U$ define $\alpha^m_{x,y}$ to be the m-block defined by $\alpha^m_{x,y}(i,j) = \alpha(mx + i, my + j)$ for $0 \le i, j < m$.

For a partial function $R : B^U_m \dashrightarrow T$, define the *assembly representation function* $R^* : A^U \dashrightarrow A^T$ such that $R^*(\alpha) = \beta$ if and only if $\beta(x,y) = R(\alpha^m_{x,y})$ for all $x, y \in \mathbb{Z}^2$. Further, α is said to map *cleanly* to β under R^* if either (1) for all non empty blocks $\alpha^m_{x,y}$, $(x+u, y+v) \in \text{dom } \beta$ for some $u, v \in \{-1, 0, 1\}$ such that $u^2 + v^2 < 2$, or (2) α has at most one non-empty m-block $\alpha^m_{x,y}$. In other words, we allow for the existence of simulator "fuzz" directly north, south, east or west of a simulator macrotile, but we exclude the possibility of diagonal fuzz.

For a given *assembly representation function* R^*, define the *supertile representation function* $\tilde{R} : \tilde{A}^U \dashrightarrow \mathcal{P}(A^T)$ such that $\tilde{R}(\tilde{\alpha}) = \{R^*(\alpha) | \alpha \in \tilde{\alpha}\}$. $\tilde{\alpha}$ is said to *map cleanly* to $\tilde{R}(\tilde{\alpha})$ if $\tilde{R}(\tilde{\alpha}) \in \tilde{A}^T$ and α maps cleanly to $R^*(\alpha)$ for all $\alpha \in \tilde{\alpha}$.

In the following definitions, let $T = (T, S, \tau)$ be a 2HAM TAS and, for some initial configuration S_T, that depends on T, let $U = (U, S_T, \tau')$ be a 2HAM TAS, and let R be an m-block representation function $R : B^U_m \dashrightarrow T$.

Definition 1. *We say that U and T have equivalent productions (at scale factor m), and we write $U \Leftrightarrow_R T$ if the following conditions hold:*

1. $\left\{ \tilde{R}(\tilde{\alpha}) | \tilde{\alpha} \in A[U] \right\} = A[T]$.

2. $\left\{ \tilde{R}(\tilde{\alpha}) | \tilde{\alpha} \in A_\square[U] \right\} = A_\square[T]$.

3. *For all $\tilde{\alpha} \in A[U]$, $\tilde{\alpha}$ maps cleanly to $\tilde{R}(\tilde{\alpha})$*

Equivalent production tells us that a simulating system U produces exactly the same set of assemblies as the simulated system T, modulo scale factor (with the representation function providing the mapping of assemblies between systems). While this is a powerful set of conditions ensuring that the simulator makes the same assemblies, it does not provide a guarantee that the simulator makes them in the *same way*. Namely, we desire a simulator to make the same assemblies, but also by following the same assembly sequences (again modulo scale and application of the representation function). We call this the *dynamics* of the systems and capture the necessary equivalence in the next few definitions. It is notable that the conditions required for the dynamics of the systems to be equivalent, *following* and *modeling*, are strong enough that equivalent production follows in a straightforward way from them, and therefore is redundant. However, we include it for completeness and clarity. We require some notation pertaining to assembly sequences.

For two supertiles $\tilde{\alpha}$ and $\tilde{\beta}$, and temperature $\tau \in \mathbb{N}$, define the *combination set* $C^{\tau}_{\tilde{\alpha},\tilde{\beta}}$ to be the set of all supertiles $\tilde{\gamma}$ such that there exist $\alpha \in \tilde{\alpha}$ and $\beta \in \tilde{\beta}$ such that (1) α and β are disjoint (steric protection), (2) $\gamma \equiv \alpha \cup \beta$ is τ-stable, and (3) $\gamma \in \tilde{\gamma}$. That is, $C^{\tau}_{\tilde{\alpha},\tilde{\beta}}$ is the set of all τ-stable supertiles that can be obtained by "attaching" $\tilde{\alpha}$ to $\tilde{\beta}$ stably, with $|C^{\tau}_{\tilde{\alpha},\tilde{\beta}}| > 1$ if there is more than one position at which β could attach stably to α.

Given a TAS $\mathcal{T} = (T, S, \tau)$, define an *assembly sequence* of $\mathcal{T}$ to be a sequence of states $\boldsymbol{S} = (S_i \mid 0 \leq i < k)$ (where $k = \infty$ if $\boldsymbol{S}$ is an infinite assembly sequence), and S_{i+1} is constrained based on S_i in the following way: There exist supertiles $\tilde{\alpha}, \tilde{\beta}, \tilde{\gamma}$ such that (1) $\tilde{\gamma} \in C^{\tau}_{\tilde{\alpha},\tilde{\beta}}$, (2) $S_{i+1}(\tilde{\gamma}) = S_i(\tilde{\gamma}) + 1$,[1] (3) if $\tilde{\alpha} \neq \tilde{\beta}$, then $S_{i+1}(\tilde{\alpha}) = S_i(\tilde{\alpha}) - 1$, $S_{i+1}(\tilde{\beta}) = S_i(\tilde{\beta}) - 1$, otherwise if $\tilde{\alpha} = \tilde{\beta}$, then $S_{i+1}(\tilde{\alpha}) = S_i(\tilde{\alpha}) - 2$, and (4) $S_{i+1}(\tilde{\omega}) = S_i(\tilde{\omega})$ for all $\tilde{\omega} \notin \{\tilde{\alpha}, \tilde{\beta}, \tilde{\gamma}\}$. That is, S_{i+1} is obtained from S_i by picking two supertiles from S_i that can attach to each other, and attaching them, thereby decreasing the count of the two reactant supertiles and increasing the count of the product supertile.

The *result* of a supertile assembly sequence $\vec{\alpha}$ is the unique supertile $\mathrm{res}(\vec{\alpha})$ such that there exist an assembly $\alpha \in \mathrm{res}(\vec{\alpha})$ and, for each $0 \leq i < k$, assemblies $\alpha_i \in \tilde{\alpha}_i$ such that $\mathrm{dom}\,\alpha = \bigcup_{0 \leq i < k} \mathrm{dom}\,\alpha_i$ and, for each $0 \leq i < k$, $\alpha_i \sqsubseteq \alpha$. For all supertiles $\tilde{\alpha}, \tilde{\beta}$, we write $\tilde{\alpha} \rightarrow_{\mathcal{T}} \tilde{\beta}$ (or $\tilde{\alpha} \rightarrow \tilde{\beta}$ when $\mathcal{T}$ is clear from context) to denote that there is a supertile assembly sequence $\vec{\alpha} = (\tilde{\alpha}_i \mid 0 \leq i < k)$ such that $\tilde{\alpha}_0 = \tilde{\alpha}$ and $\mathrm{res}(\vec{\alpha}) = \tilde{\beta}$. We write $\tilde{\alpha} \rightarrow^1_{\mathcal{T}} \tilde{\beta}$ ($\tilde{\alpha} \rightarrow^1 \tilde{\beta}$) to denote an assembly sequence of length 1 from $\tilde{\alpha}$ to $\tilde{\beta}$ and $\tilde{\alpha} \rightarrow^{\leq 1}_{\mathcal{T}} \tilde{\beta}$ ($\tilde{\alpha} \rightarrow^{\leq 1} \tilde{\beta}$) to denote an assembly sequence of length 1 from $\tilde{\alpha}$ to $\tilde{\beta}$ if $\tilde{\alpha} \neq \tilde{\beta}$ and an assembly sequence of length 0 otherwise.

Definition 2. *We say that $\mathcal{T}$ follows $\mathcal{U}$ (at scale factor m), and we write $\mathcal{T} \dashv_R \mathcal{U}$ if, for any $\tilde{\alpha}, \tilde{\beta} \in A[\mathcal{U}]$ such that $\tilde{\alpha} \rightarrow^1_{\mathcal{U}} \tilde{\beta}$, $\tilde{R}(\tilde{\alpha}) \rightarrow^{\leq 1}_{\mathcal{T}} \tilde{R}\left(\tilde{\beta}\right)$.*

Definition 3. *We say that $\mathcal{U}$ weakly models $\mathcal{T}$ (at scale factor m), and we write $\mathcal{U} \models_R \mathcal{T}$ if, for any $\tilde{\alpha}, \tilde{\beta} \in A[\mathcal{T}]$ such that $\tilde{\alpha} \rightarrow^1_{\mathcal{T}} \tilde{\beta}$, for all $\tilde{\alpha}' \in A[\mathcal{U}]$ such that $\tilde{R}(\tilde{\alpha}') = \tilde{\alpha}$, there exists an $\tilde{\alpha}'' \in A[\mathcal{U}]$ such that $\tilde{R}(\tilde{\alpha}'') = \tilde{\alpha}$, $\tilde{\alpha}' \rightarrow_{\mathcal{U}} \tilde{\alpha}''$, and $\tilde{\alpha}'' \rightarrow^1_{\mathcal{U}} \tilde{\beta}'$ for some $\tilde{\beta}' \in A[\mathcal{U}]$ with $\tilde{R}\left(\tilde{\beta}'\right) = \tilde{\beta}$.*

Definition 4. *We say that $\mathcal{U}$ strongly models $\mathcal{T}$ (at scale factor m), and we write $\mathcal{U} \models^+_R \mathcal{T}$ if for any $\tilde{\alpha}, \tilde{\beta} \in A[\mathcal{T}]$ such that $\tilde{\gamma} \in C^{\tau}_{\tilde{\alpha},\tilde{\beta}}$, then for all $\tilde{\alpha}', \tilde{\beta}' \in A[\mathcal{U}]$ such that $\tilde{R}(\tilde{\alpha}') = \tilde{\alpha}$ and $\tilde{R}\left(\tilde{\beta}'\right) = \tilde{\beta}$, it must be that there exist $\tilde{\alpha}'', \tilde{\beta}'', \tilde{\gamma}' \in A[\mathcal{U}]$, such that $\tilde{\alpha}' \rightarrow_{\mathcal{U}} \tilde{\alpha}''$, $\tilde{\beta}' \rightarrow_{\mathcal{U}} \tilde{\beta}''$, $\tilde{R}(\tilde{\alpha}'') = \tilde{\alpha}$, $\tilde{R}\left(\tilde{\beta}''\right) = \tilde{\beta}$, $\tilde{R}(\tilde{\gamma}') = \tilde{\gamma}$, and $\tilde{\gamma}' \in C^{\tau'}_{\tilde{\alpha}'',\tilde{\beta}''}$.*

---

[1] with the convention that $\infty = \infty + 1 = \infty - 1$.

Definition 5. *Let* $\mathcal{U} \Leftrightarrow_R \mathcal{T}$ *and* $\mathcal{T} \dashv_R \mathcal{U}$.

1. $\mathcal{U}$ *simulates* $\mathcal{T}$ *(at scale factor m) if* $\mathcal{U} \models_R^- \mathcal{T}$.
2. $\mathcal{U}$ *strongly simulates* $\mathcal{T}$ *(at scale factor m) if* $\mathcal{U} \models_R^+ \mathcal{T}$.

For simulation, we require that when a simulated supertile $\tilde{\alpha}$ may grow, via one combination attachment, into a second supertile $\tilde{\beta}$, then any simulator supertile that maps to $\tilde{\alpha}$ must also grow into a simulator supertile that maps to $\tilde{\beta}$. The converse should also be true. For strong simulation, in addition to requiring that all supertiles mapping to $\tilde{\alpha}$ must be capable of growing into a supertile mapping to $\tilde{\beta}$ when $\tilde{\alpha}$ can grow into $\tilde{\beta}$ in the simulated system, we further require that this growth can take place by the attachment of *any* supertile mapping to $\tilde{\gamma}$, where $\tilde{\gamma}$ is the supertile that attaches to $\tilde{\alpha}$ to get $\tilde{\beta}$.

Note that, by these definitions, strong simulation implies simulation. That is, if system $\mathcal{T}_1$ strongly simulates $\mathcal{T}_2$, then it also simulates $\mathcal{T}_2$.

2.3 Intrinsic Universality

Let REPR denote the set of all m-block (or macrotile) representation functions. Let $\mathfrak{C}$ be a class of tile assembly systems, and let U be a tile set. We say U is *intrinsically universal* for $\mathfrak{C}$ if there are computable functions $\mathcal{R} : \mathfrak{C} \to$ REPR and $\mathcal{S} : \mathfrak{C} \to \left(A_{<\infty}^U \to \mathbb{N} \cup \{\infty\}\right)$, and a $\tau' \in \mathbb{Z}^+$ such that, for each $\mathcal{T} = (T, S, \tau) \in \mathfrak{C}$, there is a constant $m \in \mathbb{N}$ such that, letting $R = \mathcal{R}(\mathcal{T})$, $S_\mathcal{T} = \mathcal{S}(\mathcal{T})$, and $\mathcal{U}_\mathcal{T} = (U, S_\mathcal{T}, \tau')$, $\mathcal{U}_\mathcal{T}$ simulates $\mathcal{T}$ at scale m and using macrotile representation function R. That is, $\mathcal{R}(\mathcal{T})$ gives a representation function R that interprets macrotiles (or m-blocks) of $\mathcal{U}_\mathcal{T}$ as assemblies of $\mathcal{T}$, and $\mathcal{S}(\mathcal{T})$ gives the initial state used to create the necessary macrotiles from U to represent $\mathcal{T}$ subject to the constraint that no macrotile in $S_\mathcal{T}$ can be larger than a single $m \times m$ square.

3 Uniform Mappings

In this section, we define *uniform mapping* and *almost linear* uniform mapping, which will provide the basis for our results related to strong simulation. We then prove a set of facts about pairs of temperatures and these mappings, most notably that it is "easy" to find a uniform mapping between temperatures if one exists.

Definition 6. *Let* $E = \{n|n \in \mathbb{N} \text{ and } n \leq Q\}$ *and* $F = \{n|n \in \mathbb{N} \text{ and } n \leq R\}$ *for some* $Q, R \in \mathbb{Z}^+$ *with* $Q \leq R$. *Let* S *be a multiset consisting of members from* E. *Then we say that there is a* uniform mapping M *from* E *to* F *if there exists a function* $M : E \to F$ *such that* $\sum_{x \in S} M(x) \geq R$ *if and only if* $\sum_{x \in S} x \geq Q$.

We say that there is a uniform mapping from τ to τ' provided that there exists a uniform mapping from $\{1, 2, ..., \tau\}$ to $\{1, 2, ..., \tau'\}$.

Definition 7. *Let $E = \{n|n \in \mathbb{N} \text{ and } n \leq Q\}$ and $F = \{n|n \in \mathbb{N} \text{ and } n \leq R\}$ for some $Q, R \in \mathbb{Z}^+$ with $Q \leq R$, and let $M : E \to F$ be a uniform mapping from E to F. We say that M is almost linear if there exists a $c \in \mathbb{N}$ such that for all $e \in (E - \{Q\})$, $M(e) = ce$, and $M(Q) = R$.*

If a uniform mapping is almost linear, that means that other than for the greatest value in the domain of the mapping, the mapping of a number x is simply x times some constant c, where c is constant for the mapping.

Lemma 1. *There exists a uniform mapping from $E = \{1, ..., \tau\}$ to $F = \{1, ..., \tau'\}$ if and only if there exists an almost linear uniform mapping from E to F.*

Corollary 1. *For $\tau, \tau' \in \mathbb{Z}^+$ where $\tau \leq \tau'$, a uniform mapping from τ to τ' exists if and only if there exists a constant $c \in \mathbb{N}$ such that $c(\tau - 1) < \tau' \leq c\tau$.*

Corollary 2. *Let $\tau \in \mathbb{Z}^+$ and suppose that $\tau < \tau' < 2\tau - 1$ for some $\tau' \in \mathbb{Z}^+$. Then there does not exist a uniform mapping from $\{1, 2, ...\tau\}$ to $\{1, 2, ..., \tau'\}$.*

Corollary 3. *For any $\tau \in \mathbb{Z}^+$, there are a finite number of $\tau' \in \mathbb{Z}^+$ with $\tau' > \tau$ such that a uniform mapping cannot be found from τ to τ'.*

Theorem 1. *Given $\tau, \tau' \in \mathbb{Z}^+$ with $\tau \leq \tau'$, there exists an algorithm which runs in time $O(\log^2 \tau')$ and (1) determines whether or not a uniform mapping from τ to τ' exists, and (2) if so, produces that mapping.*

The following corollary will be used later in the proof of Lemma 3.

Corollary 4. *Given $\tau, \tau' \in \mathbb{N}$ such that $1 < \tau < \tau'$, if no uniform mapping exists from τ to τ', then $(\tau - 1)\lceil \frac{\tau'}{\tau} \rceil \geq \tau'$.*

4   Strong Simulation via Uniform Mappings

In this section, we provide positive results showing that for any pair of temperatures $\tau, \tau' \in \mathbb{Z}^+$ such that $\tau < \tau'$ and there is a uniform mapping from τ to τ', then there exists a tile set $U_{\tau'}$ which is intrinsically universal at temperature τ' for the class of all 2HAM systems at temperature τ.

Lemma 2. *Let $\tau, \tau' \in \mathbb{Z}^+$ with $\tau < \tau'$, such that there exists a uniform mapping M from τ to τ', and let $\mathcal{T} = (T, S, \tau)$, be an arbitrary 2HAM system at temperature τ. Then, there exists $\mathcal{T}' = (T', S', \tau')$ such that $\mathcal{T}'$ strongly simulates $\mathcal{T}$.*

To prove Lemma 2, we show how to create $\mathcal{T}'$ from $\mathcal{T}$ by using the mapping M. $\mathcal{T}'$ is essentially identical to $\mathcal{T}$, but for each glue g on a tile in T, if its strength is given by the function $\mathtt{str}(g)$, then the strength of that glue in T' is equal to $M(\mathtt{str}(g))$. Due to the properties of a uniform mapping, we show that if and only if a multiset of glues on a pair of supertiles over $\mathcal{T}$ allow those supertiles to bind in $\mathcal{T}$, the mapped glues over supertiles in $\mathcal{T}'$ will allow the equivalent supertiles in $\mathcal{T}'$ to bind. Thus, $\mathcal{T}'$ will correctly strongly simulate $\mathcal{T}$.

Lemma 2 shows that as long as there is a uniform mapping between two temperatures, for each system at the lower temperature there exists a system at the higher temperature which can strongly simulate it. Furthermore, Corollaries 2 and 3 show us that there are only a very few temperatures greater than a given τ for which a uniform mapping does not exist. Theorem 1 tells us that we can efficiently find a uniform mapping M if one exists, and by the proof of Lemma 2 we can also see that the generation of the simulating system merely requires M and time linear in the size of the system to be simulated. We now show that such a strongly simulating system can be created for a tile set which is intrinsically universal for systems at τ, resulting in a tile set which is IU for systems at temperature τ while strongly simulating them at τ'.

Theorem 2. *Let $\tau, \tau' \in \mathbb{Z}^+$ with $1 < \tau < \tau'$, such that there exists a uniform mapping M from τ to τ'. Then there exists a tile set $U_{\tau'}$ which is intrinsically universal for the class of all 2HAM systems at temperature τ, such that the simulating systems using $U_{\tau'}$ are at temperature τ'.*

The proof of Theorem 2 simply makes use of the result of [8] showing that for the class of systems at each temperature $\tau \geq 2$, there exists a tile set which is IU for that class. That IU tile set simulates at temperature τ, so we use Lemma 2 to show that for $\tau' > \tau$ where a uniform mapping exists from τ to τ', we can make a strongly simulating tile set at temperature τ' for the tile set which is IU for τ systems.

Note that the results of [8] provide for a variety of tile sets for each $\tau > 1$ such that each is IU for that τ. These tile sets provide for a variety of tradeoffs in scale factor, tile set size, and number of seed assemblies. Any such tile set U_τ can be used to create the tile set $U_{\tau'}$ from Theorem 2 to achieve the same tradeoffs since the simulation of U_τ by $U_{\tau'}$ is at scale factor 1 and there is a bijective mapping of tile types from $U_{\tau'}$ to whichever U_τ is chosen. Furthermore, an IU tile set at temperature τ can be chosen which is IU in terms of either strong simulation or standard simulation, and by those definitions the result still holds.

5   Impossibility of Strong Simulation at Higher Temperatures

Intuitively, it may appear that the class of systems at higher temperatures is more "powerful" than the class of systems at lower temperatures. In this section, we show that this is not strictly the case. Here we present a sketch of the proof by giving an example of a tile set U such that there exists a 2HAM TAS $\mathcal{T} = (T, S, 3)$ such that for any initial configuration $S_{\mathcal{T}}$ over U, the 2HAM TAS $\mathcal{U} = (U, S_{\mathcal{T}}, 4)$ does not strongly simulate $\mathcal{T}$. This gives an intuitive idea of the general proof which can be found in [14].

Theorem 3. *Let $\tau, \tau' \in \mathbb{N}$ be such that (1) $2 < \tau < \tau'$ and (2) there does not exist a uniform mapping from τ to τ'. For every tile set U, there exists a 2HAM TAS $\mathcal{T} = (T, S, \tau)$ such that for any initial configuration $S_{\mathcal{T}}$ over U, the 2HAM TAS $\mathcal{U} = (U, S_{\mathcal{T}}, \tau')$ does not strongly simulate $\mathcal{T}$.*

Fig. 1. The tile set for the proof of Theorem 3. Black rectangles represent strength-τ glues (labeled 1–8), and black squares represent the strength-1 glue (labeled 0).

Proof: As in [8], the idea behind this proof is to use Definitions 2 and 4 in order to show two producible supertiles in $\mathcal{T}$ which cannot bind due to insufficient strength, but whose simulating supertiles in $\mathcal{U}$ can combine. This will contradict the definition of simulation. A large part of the terminology and notation in this proof are borrowed from [8].

Our proof is by contradiction. Therefore, suppose, for the sake of obtaining a contradiction, that there exists an intrinsically universal tile set U such that, for any 2HAM TAS $\mathcal{T} = (T, S, \tau)$, there exists an initial configuration $S_\mathcal{T}$ and $\tau' \geq \tau$, such that $\mathcal{U} = (U, S_\mathcal{T}, \tau')$ strongly simulates $\mathcal{T}$ and there does not exist a uniform mapping from τ to τ'. Define $\mathcal{T} = (T, \tau)$ where T is the tile set defined in Fig. 1, the default initial state is used, and $\tau > 2$. Let $\mathcal{U} = (U, S_\mathcal{T}, \tau')$ be the temperature $\tau' \geq \tau$ 2HAM system, which uses tile set U and initial configuration $S_\mathcal{T}$ (depending on $\mathcal{T}$) to strongly simulate $\mathcal{T}$ at scale factor m. Let $\tilde{R}$ denote the supertile representation function that testifies to the fact that $\mathcal{U}$ strongly simulates $\mathcal{T}$.

We say that a supertile $\tilde{l} \in \mathcal{A}[\mathcal{T}]$ is a d-rung *left half-ladder* of height $h \in \mathbb{N}$ if it contains h tiles of the type A_2 and $h - 1$ tiles of type A_3, arranged in a vertical column, plus d tiles each of the types A_1 and A_0 for $d \in \mathbb{N}$. (An example of a τ-rung left half-ladder is shown on the left in Fig. 2a. The dotted lines show positions at which tiles of type A_1 and A_0 could potentially attach, but since a τ-rung half-ladder has exactly τ of each, only τ such locations have tiles.) Essentially, a d-rung left half-ladder consists of a single-tile-wide vertical column of height $2h - 1$ with an A_2 tile at the bottom and top, and those in between alternating between A_2, A_3, and A_4 tiles. To the east of exactly d of the A_2 tiles an A_1 tile is attached and to the east of each A_1 tile an A_0 tile is attached. These A_1-A_0 pairs, collectively, form the τ *rungs* of the left half-ladder. We enumerate the A_2 tiles appearing in $\tilde{l}$ from north to south and denote the ith A_2 tile by $A_{2,i}$. Thus, $A_{2,0}$ denotes the northernmost A_2 tile in $\tilde{l}$ and $A_{2,(d-1)}$ denotes the southernmost tile in $\tilde{l}$. We can define d-rung *right half-ladders* similarly. A d-rung *right half-ladder* of height h is defined exactly the same way but using the tile types B_3, B_2, B_1, and B_0 and with rungs growing to the left of the vertical column. The east glue of A_0 is a strength-1 glue matching the west glue of B_0.

We say that a supertile consisting only of tiles of type A_2, A_3, and A_4 is a *left bar* provided that the northernmost tile in the supertile is A_4 and the southernmost tile in the supertile is A_3. The height of a bar is the number of A_2 tiles appearing in the bar. We define a *right bar* similarly. In the case where $\tau = 3$ and $\tau' = 4$, note that there does not exist a uniform mapping from τ to τ'. Also, in this case, Fig. 2 shows the main idea of the proof of Theorem 3.

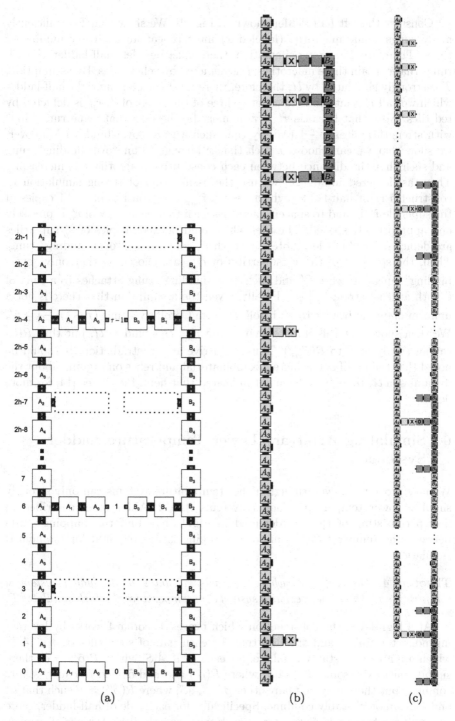

Fig. 2. (a) gives an example half-ladders with τ rungs. The squares in (b) and (c) depict macrotiles which assemble in $\mathcal{U}$ and simulate tiles $\mathcal{T}$ when $\tau' = 4$ and $\tau = 3$.

Consider the left half-ladder shown in Fig. 2b. We show that for sufficiently many rungs, some macrotile (labeled x) must repeat an arbitrary number of times. Therefore, for strong simulation, there must be a left half-ladder, $\tilde{l}'$, with rungs that contain these macrotiles. $\tilde{l}'$ is depicted by yellow tiles. By assumption, $\mathcal{T}$ is strongly simulated by $\mathcal{U}$, therefore, there must be a 3 rung right half-ladder which we call r_p' that binds to exactly three of the rungs of $\tilde{l}'$. $\tilde{r}_p'$ is depicted by red tiles. Note that because $\tau' > \tau$, it must be the case that some rung binds with strength at least $\lceil \frac{\tau'}{\tau} \rceil$ (we say that such a rung "over-binds".) Moreover, we show that we can choose x such that x belongs to an "over-binding" rung and such that the distance between each consecutive macrotile x is increasing. Then, as depicted in Fig. 2c, we use the assumption of strong simulation to construct a right half-ladder which we call $\tilde{r}_{bar}'$ that consists of $\tau - 1$ copies of the supertile $\tilde{r}_p'$ bound to spacer macrotiles such that each copy of $\tilde{r}_p'$ is precisely and appropriately spaced. The tiles which bind between copies of $\tilde{r}_p'$ supertiles are depicted by blue tiles. Note that each $\tilde{r}_p'$ contains an "over-binding" rung. Then, the spacings of the $\tilde{r}_p'$ supertiles of $\tilde{r}_{bar}'$ are chosen so that only "over-binding" rungs attach to $\tilde{l}'$ and each "over-binding" rung attaches to a rung of $\tilde{l}'$ with at least strength $\lceil \frac{\tau'}{\tau} \rceil$. Finally, given the assumption that there is not a uniform mapping from τ to τ', it follows from Corollary 4 that $(\tau - 1)\lceil \frac{\tau'}{\tau} \rceil \geq \tau'$. We then show that this implies that $\tilde{l}'$ and $\tilde{r}_{bar}'$ can bind in $\mathcal{U}$, but that $\tilde{R}(\tilde{l}')$ cannot stably bind to $\tilde{R}(\tilde{r}_{bar}')$. Thus, we arrive at a contradiction. It should be noted that the proof is not merely combinatorial and relies on arguing about the dynamics of $\mathcal{U}$, though we have not indicated that here. Please see [14] for more detail.

6   Simulating Arbitrary Lower Temperature Ladder Systems

We now prove that, even though higher temperature systems can only strongly simulate lower temperature ladder systems (the 2HAM system described in Sect. 5 consisting of the tile depicted in Fig. 1) if a uniform mapping exists between the temperatures, a uniform mapping is not required for (standard) simulation.

Theorem 4. *For $\tau, \tau' \in \mathbb{N}$ where $1 < \tau < \tau'$, let $\mathcal{T}$ be the ladder system at temperature τ. Then, there exists a system $\mathcal{S}$ at temperature τ' which simulates $\mathcal{T}$.*

At a high-level, the construction which proves Theorem 4 works by leveraging nondeterminism and the fact that for each pair of supertiles $\tilde{\alpha}, \tilde{\beta} \in \mathcal{A}[\mathcal{T}]$ which are able to τ-stably combine, for each $\tilde{\alpha}' \in \mathcal{A}[\mathcal{S}]$ where $\tilde{R}(\tilde{\alpha}') = \tilde{\alpha}$, there simply must exist some $\tilde{\beta}' \in \mathcal{A}[\mathcal{S}]$ where $\tilde{R}(\tilde{\beta}') = \tilde{\beta}$ and $\tilde{\alpha}'$ and $\tilde{\beta}'$ can τ'-stably combine, but there may be many other $\tilde{\beta}'' \in \mathcal{A}[\mathcal{S}]$ where $\tilde{R}(\tilde{\beta}'') = \tilde{\beta}$ such that $\tilde{\alpha}'$ and $\tilde{\beta}''$ cannot τ'-stably combine. Specifically, for each side of half-ladder, there are multiple types which can form, each with exactly 0 or 1 "special" rungs.

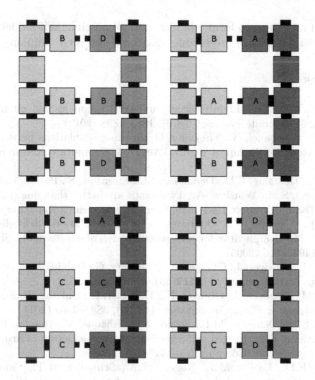

Fig. 3. Intuitive sketch of the set of half-ladders possible in the high temperature system S which simulates a low temperature ladder system T, shown without scaling. Yellow: B-type half-ladders, Blue: C-type half-ladders, Red: A-type half-ladders, Green: D-type half-ladders. Each type of half-ladder is shown once with no special rung and once with one special rung (the most possible), and each is paired with the type of half-ladder with which it could bind if each had at least τ rungs in matching locations (after translating appropriately). Note that the spacing and ordering of rungs can be arbitrary, and also that spacing tiles are left out for compactness, so rungs are closer together and shorter than they would actually be. All pairs of rungs of different types bind with each other with strength 1 (due to the H glues on their bottom tiles - not shown), and all pairs of rungs of the same type bind with strength $\tau' - \tau + 1$ due to the sum of the H glue (strength 1) and "type" glue (strength $\tau' - \tau$) bindings (Color figue online).

(See Fig. 3 for a schematic example.) All rungs on a left half-ladder can combine with all rungs on a right half-ladder with strength 1, but whenever rungs of the same type combine, they do so with strength $\tau' - \tau + 1$. The formation of all half-ladder supertiles guarantees that any pair of oppositely facing half-ladders can have no more than one pair of rungs with matching types, and for each half-ladder with τ or more rungs there exists a producible oppositely facing half-ladder with rungs in matching locations and one of them matching in type. (Note that S simulates at scale factor 2.) In such a way, τ rungs in matching locations of two oppositely facing half-ladders all guaranteed to be sufficient and necessary to form a ladder, and all possible half-ladder and ladder representing supertiles are producible, making S correctly simulate T.

References

1. Arrighi, P., Grattage, J.: Intrinsically universal n-dimensional quantum cellular automata. J. Comput. Syst. Sci. **78**(6), 1883–1898 (2012)
2. Arrighi, P., Schabanel, N., Theyssier, G.: Intrinsic simulations between stochastic cellular automata. Technical report 1208.2763, Computing Research Repository (2012)
3. Cannon, S., Demaine, E.D., Demaine, M.L., Eisenstat, S., Patitz, M.J., Schweller, R., Summers, S.M., Winslow, A.: Two hands are better than one (up to constant factors). Technical report 1201.1650, Computing Research Repository (2012)
4. Cheng, Q., Aggarwal, G., Goldwasser, M.H., Kao, M.-Y., Schweller, R.T., de Espanés, P.M.: Complexities for generalized models of self-assembly. SIAM J. Comput. **34**, 1493–1515 (2005)
5. Delorme, M., Mazoyer, J., Ollinger, N., Theyssier, G.: Bulking I: an abstract theory of bulking. Theo. Comput. Sci. **412**(30), 3866–3880 (2011)
6. Delorme, M., Mazoyer, J., Ollinger, N., Theyssier, G.: Bulking II: classifications of cellular automata. Theo. Comput. Sci. **412**(30), 3881–3905 (2011)
7. Demaine, E.D., Demaine, M.L., Fekete, S.P., Ishaque, M., Rafalin, E., Schweller, R.T., Souvaine, D.L.: Staged self-assembly: nanomanufacture of arbitrary shapes with $O(1)$ glues. Nat. Comput. **7**(3), 347–370 (2008)
8. Demaine, E.D., Patitz, M.J., Rogers, T.A., Schweller, R.T., Summers, S.M., Woods, D.: The two-handed tile assembly model is not intrinsically universal. In: Fomin, F.V., Freivalds, R., Kwiatkowska, M., Peleg, D. (eds.) ICALP 2013, Part I. LNCS, vol. 7965, pp. 400–412. Springer, Heidelberg (2013)
9. Doty, D., Lutz, J.H., Patitz, M.J., Schweller, R.T., Summers, S.M., Woods, D.: The tile assembly model is intrinsically universal. In: Proceedings of the 53rd Annual IEEE Symposium on Foundations of Computer Science, FOCS 2012, pp. 302–310 (2012)
10. Durand, B., Róka, Z.: The game of life: universality revisited. In: Delorme, M., Mazoyer, J. (eds.) Cellular Automata. Kluwer, Alphen aan den Rijn (1999)
11. Ch, E.H., Meunier, P.-E., Rapaport, I., Theyssier, G.: Communication complexity and intrinsic universality in cellular automata. Theo. Comput. Sci. **412**(1–2), 2–21 (2011)
12. Hendricks, J., Padilla, J.E., Patitz, M.J., Rogers, T.A.: Signal transmission across tile assemblies: 3D static tiles simulate active self-assembly by 2D signal-passing tiles. In: Soloveichik, D., Yurke, B. (eds.) DNA 2013. LNCS, vol. 8141, pp. 90–104. Springer, Heidelberg (2013)
13. Hendricks, J., Patitz, M.J.: On the equivalence of cellular automata and the tile assembly model. In: Neary, T., Cook, M. (eds.) Proceedings Machines, Computations and Universality 2013, vol. 128, pp. 167–189. Open Publishing Association, New York (2013)
14. Hendricks, J., Patitz, M.J., Rogers, T.A.: The simulation powers and limitations of higher temperature hierarchical self-assembly systems. CoRR, abs/1503.04502 (2015)
15. Mazoyer, J., Rapaport, I.: Inducing an order on cellular automata by a grouping operation. In: Meinel, C., Morvan, M. (eds.) STACS 1998. LNCS, vol. 1373. Springer, Heidelberg (1998)

16. Meunier, P.E., Patitz, M.J., Summers, S.M., Theyssier, G., Winslow, A., Woods, D.: Intrinsic universality in tile self-assembly requires cooperation. In: Proceedings of the ACM-SIAM Symposium on Discrete Algorithms (SODA 2014), (Portland, OR, USA, January 5-7, 2014), pp. 752–771 (2014)
17. Ollinger, N.: Intrinsically universal cellular automata. In: The Complexity of Simple Programs, in Electronic Proceedings in Theoretical Computer Science, vol. 1, pp. 199–204 (2008)
18. Ollinger, N., Richard, G.: Four states are enough!. Theo. Comput. Sci. **412**(1), 22–32 (2011)
19. Winfree, E., Algorithmic self-assembly of DNA. Ph.D. thesis, California Institute of Technology (June 1998)
20. Woods, D.: Intrinsic universality and the computational power of self-assembly. In: MCU: Proceedings of Machines, Computations and Universality, vol. 128, pp. 16–22, University of Zürich, Switzerland. Open Publishing Association, 9–12 September 2013. doi:10.4204/EPTCS.128.5

A Characterization of NP Within Interval-Valued Computing

Benedek Nagy[1,2] and Sándor Vályi[3]([✉])

[1] Department of Mathematics, Faculty of Arts and Sciences,
Eastern Mediterranean University,
Famagusta, North Cyprus, Mersin-10, Turkey
nbenedek.inf@gmail.com
[2] Department of Computer Science, Faculty of Informatics,
University of Debrecen, Debrecen, Hungary
[3] Institute of Mathematics and Informatics, College of Nyíregyháza,
Nyíregyháza, Hungary
valyis@nyf.hu

Abstract. In this paper, a syntactic subclass of polynomial size interval-valued computations is given that characterizes NP, that is, exactly languages with non-deterministically polynomial time complexity can be decided by interval-valued computations of this subclass. This subclass refrains from using product and shift operators aside from a starting section of the computation.

Keywords: Unconventional computing · Interval-valued computing · Complexity · NP · coNP · Deterministic computing

1 Introduction

Computations are important parts of our lives. Nowadays, computers are used everywhere and algorithms (programs) are executed not only on computers, but on almost all types of electronic devices. For efficient usage of these devices, we need efficient algorithms. Thus, the theory of computations is also an important field. Several types of abstract computing devices are developed, described and analyzed, such as Turing machines [2]. Turing machines play an essential role in the theory of computing, by the Church–Turing thesis they are believed to be a universal model of computation. The first question about a new abstract model of computation is its universality, i.e., whether all the functions/languages can be computed/recognized by this new model as with Turing machines. There are several universal models known, e.g., Markov normal algorithms, generative grammars. Various non universal models are known and used as well, e.g., finite automata, pushdown automata. Unfortunately, with universal models of computing one needs to face some undecidable problems, e.g., the halting problem of Turing machines. Regarding a restricted class of problems, the recursive problems/languages (or partially recursive functions), their each instance can be solved, i.e., there is a Turing machine that halts for any input of such a problem and gives the correct answer. Then, an important question is with which

© Springer International Publishing Switzerland 2015
J. Durand-Lose and B. Nagy (Eds.): MCU 2015, LNCS 9288, pp. 164–179, 2015.
DOI: 10.1007/978-3-319-23111-2_11

complexity one can obtain a solution, or decide if a particular word (instance) belongs to the language. Various classes of problems are known depending on the complexity of the algorithms that provide solutions/decisions. Well known complexity classes are P, NP, $PSPACE$ etc. For readers who are not familiar with basic notions of complexity theory, the book [13] is advised.

The following part of the introduction is broken to two subsections, in the next subsection some motivations are given for new computational paradigms. Then a brief summary is provided about the interval-valued computing paradigm including motivations and connections to other paradigms.

1.1   Computing Paradigms

Various paradigms of computations are known, the traditional Turing machine and the classical von Neumann type computers are/were widely used. However, the nature of computation is changing, in the past 10 years it tends to the direction of parallel computing. Actually, a fixed size of parallelism was present at von Neumann architectures, the ALU used a fixed number of bits, and logical operations are executed in parallel on the bits of a byte. The size of this byte, i.e., the number of bits that are used together during a computation step was a measure of the architecture. While 8-bit computers were widespread in the 1980's, nowadays 64-bit computers are standards.

There are plenty of intractable problems known, and we cannot solve their large instances by using the classical paradigm. One of the main aims and promises of the new paradigms is that they could break or push out the border of problems that can be solved efficiently. Of course to do this, these new paradigms throw away one or more principle of Neumann type computations. Most of these models allow a kind of massive parallelism during the computation.

New, unconventional methods of computations are developed a lot in the last 20 years. In quantum computing, by entanglement, one can compute by exponential amount (on the number of used qubits) of information in a parallel manner. The major drawback(?) of these systems is that this mixed state information cannot be copied (no cloning in quantum computing). Moreover, it is collapsing by measuring it, i.e., usually, we need to have the decision by only one measurement on the system at the end of the computation, and we cannot check the partial results during the computation. Thus, we do not have (conditional) branching in our algorithms. In optical computing, the data is encoded using images and the computation is going by transforming such images. The continuous space machine, as a model belonging to this paradigm, is used in characterizations of various complexity classes [18]. At the field of bio-related computations, the DNA computing [14] and the membrane computing [15] have started and become fruitful branches of theoretical computer science, and also, they have various applications in related disciplines, e.g., in medicine and bioinformatics. One may read more about these new paradigms in [1,6,16].

Interval-valued computing is also a new, unconventional paradigm that has appeared 10 years ago. We give a brief overview in the next subsection about this paradigm also mentioning connections to other paradigms.

1.2   A Brief History of Interval-Valued Computing

Interval-valued logic, introduced by B. Nagy, is a general many valued logic that is able to graphically represent various fuzzy logical systems, including Gödel-type, Łukasiewicz-type and product logics underlining also their differences (see, e.g., [4]). In [5], a new computing paradigm, namely, interval-valued computing was introduced based on the interval-valued logic by adding shift operators to the system and showing how the work of classical computers' ALU can be simulated. This paradigm uses finite unions of subintervals of [0,1) as basic data processing units. While Turing machines are abstractions whose memory size can be arbitrarily extended in its length, this system assumes the unlimited density of data units. In [10], the notion of interval-valued computations was formalized. A computation is a sequence of operator applications on generalized intervals (interval-values). Operators are the usual Boolean ones, two shifts (left and right) and a so-called product which works as zooming. A discrete language L is decidable by (linear, polynomial, etc.) interval-valued computation if and only if there is a classical logspace algorithm that for any input word w constructs an interval-valued computation (of a size that linearly, polynomially depends on the length of w) that ends with a nonempty interval-value if and only if w is in L. This tastes like Boolean circuit computing style, the difference being that interval-valued computing works not on discrete bit sequences but on full interval-values. In the same paper we showed that QSAT (the language of true quantified Boolean formulae) can be decided by a linear interval-valued computation and that PSPACE coincides with the languages decidable by a polynomial size *restricted* interval-valued computations where restricted means that the product operator may occur only as a product by [0,1/2) which is a special starting interval-value of any computation.

The interval-valued computing has various relations to other computing paradigms. It is closely related to the classical paradigm, in the sense that the operations are somewhat similar to the classical bit-operations, but, as we will see, one can change the number of bits in the data unit dynamically during the computation. From another point of view, the interval-valued computing is similar to quantum computing in the sense that a large degree of (inner) parallelism is allowed, and there are no branching, the computation goes in a linear way (sequential and deterministic applications of the operations). The parallelism is hidden in the operations themselves, as the interval-values can be used as data units. In optical computing, transformations allowing to copy more images to a single image do a kind of compression of the information, in this way, the amount of information stored in a cell sized image of the grid, can be raised unlimitedly. In interval-valued computing the data is represented on 1-dimensional intervals, and the product operator can be used in a somewhat analogous manner, to zooming out, and thus, compressing information. Connections of 1-dimensional cellular automata and the interval-valued paradigm are presented in [7]. In [9], the connections of interval-valued systems to visual computing is demonstrated. In [11] and in [12] the notion of computable function by interval-valued computing is defined, and two functions were shown to be computable by polynomial length interval-valued computations which were interesting for the cryptography audience. The methods of these papers are relatively simple and a summarizing result as conclusion can be drawn: exactly the languages in NP can be decided by a specific type of interval-valued computations, which is characterized by the following conditions:

- the computation starts with generation of all the possible witnesses,
- and continues only with Boolean operators.

This observation is the interval-valued counterpart of the characterization of NP by languages having polynomially checkable witnesses and will be proven in this paper.

This paper is organized as follows. After some preliminaries including some important complexity classes and Boolean circuits (Sect. 2), the basic definitions of interval-valued computations are given (Sect. 3). In Sect. 4 our main theorem is stated and proven, while in Sect. 5 a brief example is shown. Finally, conclusions with some open questions close the paper.

2   Preliminaries

In this section we recall some basic concepts of the computational complexity. As source for this summary, any standard textbook on complexity theory can be considered, for example, [13] or [17].

Let Σ be a finite alphabet. Σ^* is the set of finite words over Σ. The notation $|w|$ is used for the length of a word $w \in \Sigma^*$. A language L over Σ is just a subset of Σ^*. It is decidable by a Turing machine M in polynomial time if there exist $c \in \mathbb{R}^+$ and $k \in \mathbb{N}^+$ such that for each $w \in \Sigma^*$, the run of M starting on input w takes no more steps than $c|w|^k$ and $w \in L$ if and only if M accepts w. P_Σ is the set of languages over Σ that can be decidable by a Turing machine in polynomial time. Further, P represents the class of all languages over any finite alphabet Φ being in P_Φ. Similarly, by definition, a language $L \subset \Sigma^*$ is in NP_Σ if there exists a nondeterministic Turing machine M such that there exist $c \in \mathbb{R}^+$ and $k \in \mathbb{N}^+$ that for each $w \in \Sigma^*$, all runs of M starting on input w take no more steps than $c|w|^k$ and $w \in L$ if and only if M has an accepting run on w. The notation NP is used for the class of all languages over any finite alphabet Φ being in NP_Φ. One of the most known and most challenging unsolved problems of complexity theory is to prove or disprove the class equation $P = NP$. It is highly believed that $P \neq NP$.

The witness theorem for NP states that the following two conditions are equivalent [17]:

- $L \in NP$.
- There exist polynomials p and q, and a deterministic Turing machine M, such that
  - for all x and y, the machine M runs in time $p(|x|)$ on input (y, x);
  - for all x in L, there exists a string y of length $q(|x|)$ such that M accepts (y, x);

- for all x not in L and all strings y of length $q(|x|)$, M rejects (y, x).

$coNP$ is the class of languages whose complements are in NP.

A logspace Turing machine is a Turing machine M using only a logarithmic amount of space on each of its tapes except the read-only input tape and write-only output tape. More exactly, there exist c and $d \in \mathbb{R}^+$ such that the number of used working tape cells does not exceed $d \log_2(c|w|)$, for each possible input word w. We say that an algorithm is logspace if it can be implemented by a logspace Turing machine.

In the next part, a traditional, memoryless model of computation is recalled. The Boolean circuit model of computation is given as follows. A Boolean circuit is a finite directed acyclic graph (V, E) equipped by a labeling o of the non-leaf vertices into the operator set $\{AND, OR, NOT\}$ and by a bijective labeling v of the leaf vertices onto the variable set $\{x_1, \ldots, x_n\}$ satisfying the followings:

- Any vertices have out-degree 1 except of the root vertex (it has 0 out-degree),
- Any non-leaf vertices of label AND or OR have in-degree 2, while NOT-vertices have in-degree 1.

Since a Boolean circuit in this variation has only one zero-out-degree vertex, we can use this variation only to decide languages by a yes/no answer. If more complex output is needed, then more output vertices are needed. In this paper, only these restricted versions are used and we do not use more complex versions.

Instead of graph representation, we will write a Boolean circuit over a variable set $\{x_1, \ldots, x_n\}$ in a sequential form. The sequential description of a Boolean circuit is a sequence satisfying the following conditions:

- The first n elements of the sequence are $x_1, \ldots, x_n$,
- the other elements are of form (op, i, j) where op is AND, OR or NOT and i, j are nonnegative integers less than the index of the actual element.

The meaning of this description of Boolean circuits in this variation is obvious. The output of the circuit on the input bit sequence $b_1, \ldots, b_n$ is the value of the last element of the description of the circuit assuming that input variables $x_1, \ldots, x_n$ take values of $b_1, \ldots, b_n$, respectively, and the Boolean operators work as usual. We remark that if the first element of (op, i, j) is NOT, then j is superficial, the second operand is not used.

We note here that Boolean circuits can be extended in various way, e.g., by computing not only on bits, but on natural numbers [3]. In this paper, our computations can also be viewed as a kind of extension of Boolean circuits using interval-values. In the next section we give the formal definitions of our model.

3   Interval-Valued Computations: Definitions

With the aim of keeping this paper self-contained, we recall the definitions describing interval-valued computations coined in [5]. This was formulated first in mathematical precision in [8].

An *interval-value* is a subset of interval $[0,1)$ which is a finite union of subintervals. The set of interval-values is denoted by $\mathbb{V}$.

The maximal subintervals of an interval-value are called its *components*.

Operators of interval-valued computation are Boolean set operators AND, OR, NOT, and three other: $PRODUCT$, $RSHIFT$, $LSHIFT$. NOT is a unary operator, all the others are binary. For the sake of denotational simplicity, we consider also NOT as a binary operator where the second operand is superfluous.

A *computation sequence* is a finite sequence whose first element is the constant $FIRSTHALF$ and every other element consists of a triplet (op, i, j) where op is an operator and i, j are positive integers less than the index of the actual element.

The *value* of an interval-valued computation is defined by induction of the length of the computation. Let S denote an interval-valued computation sequence, its value, denoted by $\|S\|$ is the interval-value that is obtained by the last operation of the sequence S. Let $S_{\to k}$ denote the k-length prefix of S.

First $\|FIRSTHALF\|$ is fixed to $[0, \frac{1}{2})$. If the last element of S is (op, i, j), containing a Boolean operator, then

- $\|S\| = \|S_{\to i}\| \cup \|S_{\to j}\|$ if op is OR,
- $\|S\| = \|S_{\to i}\| \cap \|S_{\to j}\|$ if op is AND,
- $\|S\| = [0, 1) \setminus \|S_{\to i}\|$ if op is NOT.

Before we define the meaning of the remaining operators, we introduce an assisting function. It returns the length of the left-most component (included maximal subinterval) of an interval-value A.

The function $Flength : \mathbb{V} \to \mathbb{R}$ can be defined as follows. If there exist $a, b \in [0, 1]$ satisfying $[a, b) \subseteq A$, $[0, a) \cap A = \emptyset$ and $[a, b') \not\subseteq A$ for all $b' \in (b, 1]$, then $Flength(A) = b - a$, otherwise $Flength(A) = 0$.

Flength helps us to define the binary shift operators on $\mathbb{V}$. The *left-shift* operator will shift the first interval-value to the left by the first-length of the second operand and remove the part which is shifted out of the interval $[0, 1)$. As opposed to this, the *right-shift* operator is defined in a circular way, i.e., the parts shifted above 1 will appear at the lower end of $[0, 1)$. In this definition we write interval-values in their "characteristic function" notation instead of the above subset notation. That is, for any $A \in \mathbb{V}$ and $x \in [0, 1]$, $A(x) = \begin{cases} 1, & \text{if } x \in A, \\ 0, & \text{otherwise.} \end{cases}$

The binary operators $Lshift$ and $Rshift$ on $\mathbb{V}$ are defined in the following way. If $x \in [0, 1]$ and $A, B \in \mathbb{V}$, then

$$Lshift(A, B)(x) = \begin{cases} A(x + Flength(B)), & \text{if } 0 \leq x + Flength(B) \leq 1, \\ 0, & \text{in other cases.} \end{cases}$$

$$Rshift(A, B)(x) = \begin{cases} A(frac(x - Flength(B))), & \text{if } x < 1, \\ 0, & \text{if } x = 1. \end{cases}$$

Here the function *frac* gives the fractional part of a real number, i.e., $frac(x) = x - \lfloor x \rfloor$, where $\lfloor x \rfloor$ is the greatest integer which is not greater than x.

Let A and B be interval-values and $x \in [0, 1)$. Then the *product* $A * B$ includes x if and only if $B(x) = 1$ and $A\left(\frac{x - x_B}{x_B - x_B}\right) = 1$, where x_B denotes the lower end-point of the B-component including x and x^B denotes the upper end-point of this component, that is, $[x_B, x^B)$ is the maximal subinterval of B containing x. The product $A * B$ is zooming out the interval-value A onto the components of B.

Now, we are ready to continue the definition of the semantics of the computation sequence. If the last element in the interval-valued computation sequence S is

$= (RSHIFT, i, j)$, then its value $\|S\|$ is $Rshift(\|S_{\to i}\|, \|S_{\to j}\|)$
and the other cases (shift to the left, product) are defined in a similar way, that is,
$= (LSHIFT, i, j)$, then its value $\|S\|$ is $Lshift(\|S_{\to i}\|, \|S_{\to j}\|)$ and
$= (PRODUCT, i, j)$, then its value $\|S\|$ is $Product(\|S_{\to i}\|, \|S_{\to j}\|)$.

We say that a language $L \subseteq \Sigma^*$ is *decidable by a linear interval-valued computation* if and only if there is a positive constant c and a logarithmic space algorithm $\mathcal{A}$ with the following properties. For each input word $w \in \Sigma^*$, $\mathcal{A}$ constructs an appropriate interval-valued computation sequence $\mathcal{A}(w)$ such that $|\mathcal{A}(w)|$ is not greater than $c|w|$ and $w \in L$ if and only if $\|\mathcal{A}(w)\|$ is nonempty.

We say that a language $L \subseteq \Sigma^*$ is *decidable by a polynomial interval-valued computation* if and only if there is a positive constant c, an integer $k \geq 0$ and a logarithmic space algorithm $\mathcal{A}$ with the following properties. For each input word $w \in \Sigma^*$, $\mathcal{A}$ constructs an appropriate interval-valued computation sequence $\mathcal{A}(w)$ such that $|\mathcal{A}(w)|$ is not greater than $c|w|^k$ and $w \in L$ if and only if $\|\mathcal{A}(w)\|$ is nonempty.

Note that only $[a, b)$ type interval components are used in these computations, where both a and b is of the form $\frac{x}{2^m}$ for some $m \geq 1$. Actually, the largest value m that is needed to express all interval-values of the computation sequence, gives a kind of resolution of the actual computation. This resolution is connected to the number of used $PRODUCT$ operations: by multiplying $FIRSTHALF$ by an interval-value, usually m is increased by 1, in this way, doubling the number of stored bits in an interval-value. This measure, the bit-hight of a computation was introduced in [10]. In this way, an interval-valued computation uses a dynamically changing amount of information in an interval-value, i.e., growing number of bits can be coded into an interval-value as the computation proceeds. We note that in other possible variations of interval-valued computing, not only this kind of interval-values can occur.

4   Results

Without significant loss of generality, it is assumed that $\Sigma = \{0, 1\}$.

Theorem 1. *For any $L \subset \Sigma^*$, condition (X) is equivalent with the condition $L \in NP_\Sigma$.*
(X): There exist $c, d > 0$, $k, p > 0$ and a logspace algorithm $\mathcal{A}$ mapping an interval-valued computation sequence $\mathcal{A}(w)$ to any word $w \in \Sigma^*$ in such a way that

(I) for each word $w \in \Sigma^*$, w belongs to L if and only if $\|\mathcal{A}(w)\|$ is nonempty,

(II) the length of $\mathcal{A}(w)$ is less than $c|w|^k$,

(III) the $(3d|w|^p + 1)$-length prefix of $\mathcal{A}(w)$ depends only on $|w|$ and is exactly the following sequence:

K_1 is $FIRSTHALF$,

K_2 is $(RSHIFT, 1, 1)$,

K_3 is $(OR, 1, 2)$ and

K_4 is $(OR, 1, 1)$.

For all positive integers $2 \le k \le d|w|^p$,

$K_{3k-1} = (PRODUCT, 1, 3k - 2)$,

$K_{3k} = (RSHIFT, 3k - 1, 3k - 2)$ and

$K_{3k+1} = (OR, 3k, 3k - 1)$.

(IV) the remaining part of $\mathcal{A}(w)$ involves only Boolean operators.

If an interval-valued computation have the given prefix, then the following statements can be established by induction on k:

Lemma 1. *For all positive integer k, if $k \le d|w|^p$, then*

$$\|K_{\to 3k+1}\| = \bigcup_{l=0}^{2^{k-1}-1} \left[\frac{2l}{2^k}, \frac{2l+1}{2^k} \right).$$

Lemma 2. *Let $n = d|w|^p$. For each bit sequence $y_1, \ldots, y_n$, with the choice*

$$r = \sum_{i=1}^{n} \frac{1 - y_i}{2^i},$$

the following holds for any $k \in \{1, \ldots, n\}$:

$$r \in \|K_{\to 3k+1}\|$$

if and only if

$$y_k = 1.$$

Now, we are ready to prove our main theorem.

Proof. Direction $(\Rightarrow)$ of Theorem 1.

If $L \in NP$ then, by the witness theorem for NP, there exist $e > 1$, $q \in \mathbb{N}^+$ and $W \subset \Sigma^* \times \Sigma^*$ such that the followings hold.

- W is in P, that is, W is decidable in polynomial time,
- for each pair $(y, w) \in W$, $|y| \le e|w|^q$ holds,
- for each $w \in \Sigma^*$, $w \in L$ if and only if $(\exists y \in \Sigma^*)(y, w) \in W$.

We set d in condition (X) to e and let $p = q$.

Since W is in P, there exists a uniform polynomial size Boolean circuit family that accepts W [13], that is, there exists a logspace algorithm $\mathcal{B}$ and there exist $f > 0$ and $t > 0$, so that from any unary encoded positive integer m, $\mathcal{B}$ can construct a description of a Boolean circuit of size $f \cdot m^t$ that accepts exactly the m-length elements of W.

Let $c = fd^t$ and let $k = tp$.

We give a logspace algorithm $\mathcal{A}$ satisfying (I)–(IV) with respect to the just defined constants c,d,k and p. The input of $\mathcal{A}$ is $w \in \Sigma^*$. $\mathcal{A}$ should respond to w by an interval-valued computation sequence. The answer begins with the given prefix with the just defined d and p, that is, the $(3d|w|^p + 1)$-length prefix of $\mathcal{A}(w)$ is exactly the prescribed sequence in (III).

Lemma 2 makes it possible to represent all the possible witnesses for w.

After that, $\mathcal{A}$ continues its work based on the bits w_i in the input word w $(i = 1, \ldots, |w|)$.

$$K_{3d|w|^p+1+i} = \begin{cases} OR(1,2), & \text{if } w_i = 1, \\ AND(1,2), & \text{in the other case.} \end{cases}$$

Next, we can establish the following statement as a part of the proof.

Lemma 3. *Let $w \in \Sigma^*$ be an input word. For every possible witness y no longer than $d|w|^p$, there is $r \in [0,1)$ such that $r \in \|K_4\| \Leftrightarrow y_1 = 1$, ..., $r \in \|K_{3d|w|^p+1}\| \Leftrightarrow y_{d|w|^p} = 1$, furthermore, $w_1 = 1 \Leftrightarrow r \in \|K_{3d|w|^p+1+1}\|, \ldots, w_{|w|} = 1 \Leftrightarrow r \in \|K_{3d|w|^p+1+|w|}\|$. Further, r can be chosen by the same formula as in Lemma 2.*

In the closing section, $\mathcal{A}$ follows the calculation of the Boolean circuit BC created by $\mathcal{B}$ to the input word $1^{d|w|^p+|w|}$ step-by-step, Boolean operator by Boolean operator. The number of input bits of BC is $d|w|^p + |w|$, $d|w|^p$ pieces for the witness and $|w|$ for the original input word. If BC accesses the k-th bit of the witness, then the interval-valued computation sequence constructed by $\mathcal{A}$ accesses operand $3k + 1$ while if BC accesses the i-th bit of the original input word w, then the sequence created by $\mathcal{A}$ accesses operand with an index $d|w|^p + 1 + i$. It is clear that if $\mathcal{B}$ works with logarithmic space, then $\mathcal{A}$ can work with the same amount of space, as well.

Direction ($\Leftarrow$) of Theorem 1.

Let us assume that there exist c, d, k, p and a logspace algorithm $\mathcal{A}$ creating an interval-valued computation sequence $\mathcal{A}(w)$ to each input word $w \in \Sigma^*$ with the four properties:

(i) $(\forall w \in \Sigma^*)(w \in L \Leftrightarrow \|\mathcal{A}(w)\|$ is nonempty)
(ii) $(\forall w \in \Sigma^*)|\mathcal{A}(w)| < c|w|^k$
(iii) $\mathcal{A}(w)$ has the prescribed prefix and
(iv) the remaining operators are Boolean operators.

Let W be the set

$$\left\{ (y, w) \;\middle|\; w \in L, y \in \Sigma^*, |y| = d|w|^p, \sum_{i=1}^{d|w|^p} \frac{1 - y_i}{2^i} \in \|\mathcal{A}(w)\| \right\}.$$

It is clear that for each $(y, w) \in W$, $|y| \le d|w|^p$ holds. It can also be shown by Lemma 1 and 2, that for each $w \in \Sigma^*$, $w \in L$ holds if and only if ($\exists y \in$

$\Sigma^*)(y, w) \in W$. If we prove that W is decidable in polynomial time, then, by the witness theorem, $L \in NP$ is also demonstrated.

$W \in P$ can be proven based on the following argument. We give a logspace algorithm $\mathcal{B}$ that constructs a Boolean circuit $\mathcal{B}_w$ of size $c|w|^k$ to every input word $w \in \Sigma^*$ such that for all $(y, w) \in \Sigma^* \times \Sigma^*$, the equivalence

$$\mathcal{B}_w//y = 1 \Leftrightarrow (y, w) \in W$$

holds, where $\mathcal{C}//x$ means the result of the computation of a Boolean circuit $\mathcal{C}$ on input x, where x is a bit sequence of appropriate size. $\mathcal{B}_w$ have input size $d|w|^p$.

$\mathcal{B}_w$ can be constructed (by $\mathcal{B}$) in the following way. It deletes the prescribed $3d|w|^p + 1$-length prefix of $\mathcal{A}(w)$ and substitutes it by accessing the $d|w|^p$ number of the negated input bits, in the order $1 - y_1, \ldots, 1 - y_{d|w|^p}$. These inputs are fixed by y. The remaining (Boolean) part of $\mathcal{A}(w)$ is transformed as follows. The Boolean operators will be kept in $\mathcal{B}_w$ but their operands will be adjusted in the appropriate way according to the index shift caused by the substitution. Also the following changes will be applied in $\mathcal{B}_w$.

- whenever $\mathcal{A}$ accesses K_1, $\mathcal{B}_w$ uses $1 - y_1$,
- whenever $\mathcal{A}$ accesses K_2, $\mathcal{B}_w$ uses y_1,
- whenever $\mathcal{A}$ accesses K_3, $\mathcal{B}_w$ uses 1,
- whenever $\mathcal{A}$ accesses K_{3i+1}, $\mathcal{B}_w$ uses $1 - y_i$, for each $i \in \{1, \ldots, d|w|^p\}$,
- whenever $\mathcal{A}$ accesses K_{3i+2}, $\mathcal{B}_w$
  uses $(1 - y_i) \wedge (1 - y_{i+1})$, for each $i \in \{1, \ldots, d|w|^p - 1\}$,
- whenever $\mathcal{A}$ accesses K_{3i+3}, $\mathcal{B}_w$
  uses $y_i \wedge (1 - y_{i+1})$, for each $i \in \{1, \ldots, d|w|^p - 1\}$.

We note that in this way, all input variables of $\mathcal{B}_w$ will be fixed to a constant depending only on input y.

If we denote the result of the i-th step of the computation sequence of a Boolean circuit $\mathcal{C}$ by $\mathcal{C}_{\to i}//x$, then we can establish the followings. Here $r(y)$ denotes

$$\sum_{i=1}^{d|w|^p} \frac{1 - y_i}{2^i}.$$

- If $i \in \{1, \ldots, d|w|^p\}$, then

$$(\mathcal{B}_w)_{\to i}//y = 1 - y_i = (r(y) \in \|K_{3i+1}\|);$$

- if $i \in \{1, \ldots, |\mathcal{A}(w)| - (3d|w|^p + 1)\}$, then

$$(\mathcal{B}_w)_{\to d|w|^p + i}//y = (r(y) \in \|\mathcal{A}(w)_{\to 3d|w|^p + 1 + i}\|).$$

The last case of the last item describes the last operation of $\mathcal{B}_w$ and shows that

$$\mathcal{B}_w//y = 1$$

if and only if

$$r(y) \in \|\mathcal{A}(w)\|,$$

in other words,

$$(y, w) \in W.$$

In $\mathcal{B}_w$, all input variables are fixed by y, so to decide $(y, w) \in W$, it is enough to evaluate this Boolean circuit. (Only evaluation, no search for truth valuations making true the output of that circuit.) It can be achieved in linear time according to the size of the circuit – and this size is polynomial according to the size (y, w) because it is built by a logspace algorithm having input size $|w|$.

We have finished both directions of the proof for Theorem 1. □

The following theorem is analogous to the previous theorem.

Theorem 2. *For any $L \subset \Sigma^*$, the following condition is equivalent to the condition $L \in coNP_\Sigma$.*
There exist $c, d > 0$, $k, p > 0$ and a logspace algorithm $\mathcal{A}$ mapping an interval-valued computation sequence $\mathcal{A}(w)$ to any word $w \in \Sigma^*$ in such a way that

– *for each word $w \in \Sigma^*$, w belongs to L if and only if $\|\mathcal{A}(w)\|$ is empty,*
– *the length of $\mathcal{A}(w)$ is less than $c|w|^k$,*
– *the $(3d|w|^p + 1)$-length prefix of $\mathcal{A}(w)$ depends only on $|w|$ and is exactly the following sequence:*
   K_1 is $FIRSTHALF$,
   K_2 is $(RSHIFT, 1, 1)$,
   K_3 is $(OR, 1, 2)$ and
   K_4 is $(OR, 1, 1)$.
   For all positive integers $2 \leq k \leq d|w|^p$,
   $K_{3k-1} = (PRODUCT, 1, 3k - 2)$,
   $K_{3k}\ \ = (RSHIFT, 3k - 1, 3k - 2)$ and
   $K_{3k+1} = (OR, 3k, 3k - 1)$.
– *the remaining part of $\mathcal{A}(w)$ involves only Boolean operators.*

Proof (Outline). $L \in coNP$ means by definition that $\overline{L} \in NP$. By a well-known theorem for coNP [17], what is analogous to the witness theorem for NP, $L \in coNP$ is equivalent to the existence of $c > 0$ and $k \in \mathbb{N}^+$ and a language $R \subseteq \Sigma^* \times \Sigma^*$ that the followings hold:

– $\forall (y, x) \in R : |y| < c|x|^k$,
– $\forall x \in \Sigma^* : (x \in L \Leftrightarrow (\forall y \in \Sigma^*)(y, x) \in R)$,
– $R \in P_\Sigma$.

Based on this characterization, the proof of Theorem 2 can be achieved in an analogous way to the proof of Theorem 1. □

5 Example

We use the Hamiltonian path problem for finite directed graphs as an example. It is a well-known NP-complete problem. A Boolean circuit family is defined that checks the binary encoded graph and that a path is Hamiltonian with respect to the given graph, then we give the details for a specific graph, including the interval-values computed by the given computation.

Let $G = (V, E)$ be a finite directed graph. Without loss of generality, $V = \{1, \ldots, n\}$ can be assumed, for a positive integer n. The graph can be encoded by its neighborhood matrix. That is, the code for G is the doubly indexed bit sequence $\langle e_{ij} \mid 1 \leq i, j \leq n \rangle$, where e_{ij} is either 0 or 1 and $e_{ij} = 1$ if and only if $(i, j) \in E$.

Let l be $\lceil \log_2(n) \rceil$. $\{0, 1\}^l$ denotes, as usual, the set of l-length binary sequences. Let $c : \{0, 1\}^l \to \{1, \ldots, 2^l\}$ be defined by

$$c(w) = 1 + \sum_{i=1}^{|w|} w_i 2^{|w|-i},$$

where c is a bijection (a shift of the decoding function of the binary number system), so its inverse is a function and also bijective. The restriction of this inverse to $\{1, \ldots, n\}$ will be denoted by c^{-1}. Then, c^{-1} is used as an encoding of vertices of the graph into l-length bit sequences.

A witness will encode a path $v_1, \ldots, v_n$ as a concatenated bit sequence

$$c^{-1}(v_1) \ldots c^{-1}(v_n).$$

We denote the bits of this nl-length sequence by a double indexing

$$\langle y_{ij} \mid 1 \leq i \leq n, 1 \leq j \leq l \rangle.$$

The formula verifying the witness works as follows. It has an input bit sequence concatenated from

$$y_{11}, \ldots, y_{1l}, \ldots, y_{n1}, \ldots, y_{nl}$$

for the witness path (each vertex on the path is encoded by l bits, $(y_{i1} \ldots y_{il}) = c^{-1}(v_i)$, for each $i = 1, \ldots, n$) and

$$e_{11}, \ldots, e_{1n}, \ldots, e_{n1}, \ldots, e_{nn}$$

for the input neighborhood matrix. The Boolean circuit can be constructed from the formula in the usual way.

The formula has the following subformulae, connected by conjunction:

(1) $(Y_i \equiv C(j)) \wedge (Y_{i+1} \equiv C(k)) \to e_{jk}$,
    where $i = 1, \ldots, n-1$, $\{j, k\} \subset \{1, \ldots, n\}$ and $j \neq k$
(2) $\neg(Y_i = Y_j)$, where $1 \leq i < j \leq n$
(3) $(Y_i \equiv C(1)) \vee (Y_i \equiv C(2)) \vee \ldots \vee (Y_i \equiv C(n))$, where $1 \leq i \leq n$

where $(Y_i = Y_j)$ is just an abbreviation for

$$(y_{i1} \leftrightarrow y_{j1}) \wedge \ldots \wedge (y_{il} \leftrightarrow y_{jl}),$$

and $(Y_i \equiv C(j))$ is also an abbreviation for

$$Z_{i1} \wedge \ldots \wedge Z_{il},$$

where $Z_{ik} = y_{ik}$ if the k-th bit of $c^{-1}(j)$ is 1 and $Z_{ik} = \neg y_{ij}$ if that bit is 0, for $k \in \{1, \ldots, l\}$.

Item (1) expresses that the vertices $(c(y_{i1} \ldots y_{il}), c(y_{(i+1)1} \ldots y_{(i+1)l}))$ constitute an edge in G and the instances of (2) together express that no vertex repeats in the sequence $c(y_{11} \ldots y_{1l}), \ldots, c(y_{n1} \ldots y_{nl})$. Item (2) requires that there is no repeating vertex in the path. Item (3) says that non-code bit sequences cannot occur as vertices.

The size of this formula is $\mathcal{O}(n^3 \log(n))$, it is subquadratic (so polynomial) according to the fact that the size of the input is n^2. Also the size of the corresponding Boolean circuit does not exceed this limit, of course.

In the example the existence of Hamiltonian path in the graph

$$G = (\{1, 2, 3\}, \{(1, 2), (2, 3), (2, 1)\})$$

is decided by an interval-valued computation. The code of G is

$$010101000,$$

$$l = \lceil \log_2(n) \rceil = 2,$$

and

$$c^{-1}(1) = 00, c^{-1}(2) = 01, c^{-1}(3) = 10.$$

The possible witnesses for a Hamiltonian path in G with n vertices have length $ln = 6$. Let $y_{11}y_{12}y_{21}y_{22}y_{31}y_{32}$ denote a possible witness.

The formula testing the witness against G is the conjunction of the following formulae. They are written in an implication- and equivalence-free form and so are suitable to direct circuit implementation.

$$y_{11} \vee y_{12} \vee y_{21} \vee \neg y_{22} \vee e_{12}$$
$$y_{11} \vee y_{12} \vee \neg y_{21} \vee y_{22} \vee e_{13}$$
$$y_{11} \vee \neg y_{12} \vee y_{21} \vee y_{22} \vee e_{21}$$
$$y_{11} \vee \neg y_{12} \vee \neg y_{21} \vee y_{22} \vee e_{23}$$
$$\neg y_{11} \vee y_{12} \vee y_{21} \vee y_{22} \vee e_{31}$$
$$\neg y_{11} \vee y_{12} \vee y_{21} \vee \neg y_{22} \vee e_{32}$$
$$y_{21} \vee y_{22} \vee y_{31} \vee \neg y_{32} \vee e_{12}$$
$$y_{21} \vee y_{22} \vee \neg y_{31} \vee y_{32} \vee e_{13}$$
$$y_{21} \vee \neg y_{22} \vee y_{31} \vee y_{32} \vee e_{21}$$
$$y_{21} \vee \neg y_{22} \vee \neg y_{31} \vee y_{32} \vee e_{23}$$
$$\neg y_{21} \vee y_{22} \vee y_{31} \vee y_{32} \vee e_{31}$$
$$\neg y_{21} \vee y_{22} \vee y_{31} \vee \neg y_{32} \vee e_{32}$$
$$(y_{11} \wedge \neg y_{21}) \vee (\neg y_{11} \wedge y_{21}) \vee (y_{12} \wedge \neg y_{22}) \vee (\neg y_{12} \wedge y_{22})$$

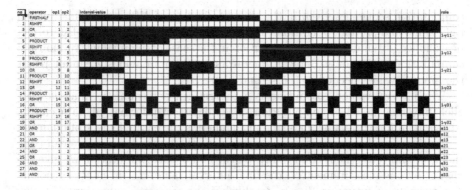

Fig. 1. First part of the interval-valued computation of the example.

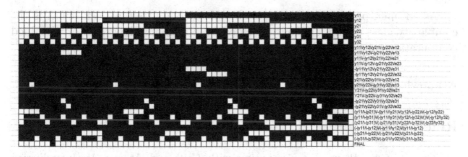

Fig. 2. Some of the interval-values in the second part of the computation of the example, including the final result.

$$(y_{11} \wedge \neg y_{31}) \vee (\neg y_{11} \wedge y_{31}) \vee (y_{12} \wedge \neg y_{32}) \vee (\neg y_{12} \wedge y_{32})$$
$$(y_{21} \wedge \neg y_{31}) \vee (\neg y_{21} \wedge y_{31}) \vee (y_{22} \wedge \neg y_{32}) \vee (\neg y_{22} \wedge y_{32})$$
$$(\neg y_{11} \wedge \neg y_{12}) \vee (\neg y_{11} \wedge y_{12}) \vee (y_{11} \wedge \neg y_{12})$$
$$(\neg y_{21} \wedge \neg y_{22}) \vee (\neg y21 \wedge y_{22}) \vee (y_{21} \wedge \neg y_{22})$$
$$(\neg y_{31} \wedge \neg y_{32}) \vee (\neg y_{31} \wedge y_{32}) \vee (y_{31} \wedge \neg y_{32})$$

In the followings, the constructed interval-valued computation is drawn in some details. In Fig. 1 the fixed but size-dependent part is displayed. In Fig. 2, however, only the listed subformulae are presented together with interval-values belonging to these formulae. The final result is nonempty according to witness $y_{11} = 0, y_{12} = 0, y_{21} = 0, y_{22} = 1, y_{31} = 1, y_{32} = 0$, that is, to path 123.

6   Conclusions, Further Remarks

Observe that left-shift operator was not used in our characterization. Moreover, according to a remark of an anonymous referee, the given characterization can be strengthened in the following sense. The first phase of the interval-value computation sequences given in the characterization of NP uses originally both right shifts and product operations, besides Boolean ones. The usage of shifts can be excluded obtaining an optimal solution in the sense of minimal instruction set: computations using only Boolean operators and shifts obviously cannot result the interval-values needed in the second phase.

The interval-valued computing is a theoretical, deterministic, massive parallel universal model of computation. Several bits of information/data can be stored in an interval-value, in a similar manner as the entanglement of quantum computing. But here we can easily manipulate the interval-values, we can copy them, we can reuse them etc. Finally, by the end of the computation we give the answer by one measurement on the final interval-value, if it is empty or not. It is a question of future studies to establish more concrete connections between interval-valued and other paradigms, e.g., quantum, optical or membrane computing.

As an open problem within interval-valued computing, we ask, what is the complexity class of the languages decided by polynomial length interval-valued computations allowing arbitrary products not only products by $\left[0, \frac{1}{2}\right)$.

Acknowledgments. Reviewers' remarks and advices are gratefully acknowledged.

References

1. Calude, C.S., Păun, G.: Computing with Cells and Atoms: An Introduction to Quantum, DNA and Membrane Computing. Taylor & Francis/Hemisphere, London, Bristol (2001)
2. Hopcroft, J.E., Ullman, J.D.: Introduction to Automata Theory, Languages, and Computation. Addison-Wesley, Reading (1979)
3. McKenzie, P., Wagner, K.W.: The complexity of membership problems for circuits over sets of natural numbers. Comput. complex. **16**(3), 211–244 (2007)
4. Nagy, B.: A general fuzzy logic using intervals, In: HUCI 2005: 6th International Symposium of Hungarian Researchers on Computational Intelligence, Budapest, Hungary, pp. 613–624 (2005)
5. Nagy, B.: An interval-valued computing device. In: CiE 2005, Computability in Europe: New Computational Paradigms, Amsterdam, Netherlands, pp. 166–177 (2005)
6. Nagy, B.: Új Számítási Paradigmák: Bevezetés az Intervallum-értékű, a DNS-, a Membrán- és a Kvantumszámítógépek elméletébe (New Computing Paradigms: Introduction to Interval-Valued, DNA, Membrane and Quantum Computing, in Hungarian). Typotex, Budapest (2014)
7. Nagy, B., Major, S.R.: Connection between interval-valued computing and cellular automata. In: CINTI 2013: 14th IEEE International Symposium on Computational Intelligence and Informatics, Budapest, Hungary, pp. 225–230 (2013)
8. Nagy, B., Vályi, S.: Solving a PSPACE-complete problem by a linear interval-valued computation. In: CiE 2006, Computability in Europe: Logical Approaches to Computational Barriers. University of Wales, Swansea, UK, pp. 216–225 (2006)
9. Nagy, B., Vályi, S.: Visual reasoning by generalized interval-values and interval temporal logic. In: CEUR Workshop Proceedings, vol. 274, pp. 13–26 (2007)

10. Nagy, B., Vályi, S.: Interval-valued computations and their connection with PSPACE. Theor. Comput. Sci. **394**, 208–222 (2008)
11. Nagy, B., Vályi, S.: Prime factorization by interval-valued computing. Publicationes Mathematicae Debrecen **79**, 539–551 (2011)
12. Nagy, B., Vályi, S.: Computing discrete logarithm by interval-valued paradigm. Electron. Proc. Theor. Comput. Sci. **143**, 76–86 (2014)
13. Papadimitriou, C.: Computational Complexity. Addison Wesley, Reading (1994)
14. Păun, Gh., Rozenberg, G., Salomaa, A.: DNA Computing: New computing paradigms. Springer, Berlin (1998)
15. Păun, G., Rozenberg, G., Salomaa, A. (eds.): Handbook of Membrane Computing. Oxford University Press, Oxford (2010)
16. Rozenberg, G., Bäck, T., Kok, J.N. (eds.): Handbook of Natural Computing. Springer, Heidelberg (2012)
17. Sipser, M.: Introduction to the Theory of Computation. Cengage Learning, Boston (2012)
18. Woods, D., Naughton, T.J.: Optical computing. Appl. Math. Comput. **215**, 1417–1430 (2009)

Universality in Infinite Petri Nets

Dmitry A. Zaitsev[(⊠)]

Faculty of Computer Science, Vistula University,
St. Stoklosy, 3, 02-787 Warsaw, Poland
d.zaitsev@vistula.edu.pl
http://member.acm.org/~daze

Abstract. Finite classical Petri nets are non-Turing-complete. Two infinite Petri nets are constructed which simulate the linear cellular automaton Rule 110 via expanding traversals of the cell array. One net is obtained via direct simulation of the cellular automaton while the other net simulates a Turing machine, which simulates the cellular automaton. They use cell models of 21 and 14 nodes, respectively, and simulate the cellular automaton in polynomial time. Based on known results we conclude that these Petri nets are Turing-complete and run in polynomial time. We employ an induction proof technique that is applicable for the formal proof of Rule 110 ether and gliders properties further to the constructs presented by Matthew Cook.

Keywords: Universal Petri net · Infinite Petri net · Linear cellular automaton · Turing machine · Simulation · Complexity

1 Introduction

Recently, small polynomial time universal extended Petri nets [22] have been constructed in the class of inhibitor multichannel Petri nets, where an inhibitor net [16] implements a check on a place to determine whether or not it has a zero marking, and a multichannel transition [20] allows the firing in a few instances at a step. Universal nets were obtained via simulation of Neary and Woods' small weakly universal Turing machine [12]. These nets simulate the linear cellular automaton Rule 110, for which universality was proven by Matthew Cook [4]. Earlier constructed inhibitor universal Petri nets run in exponential time [21], while extending them with the multichannel transitions concept allowed an efficient simulation with polynomial time complexity [22]. Another approach [8] is based on simulating small register machines of Ivan Korec [9] by inhibitor Petri nets; however, register machines are known to be exponentially slower than Turing machine [7]. Note that, an analog of multichannel transitions [20], called "exhaustive use of rules" allowed the construction of the small universal extended spiking neural P systems running in polynomial time [13].

Finite classical Petri nets are known to be non-Turing-complete [16]. Tilak Agerwala has proven Turing-completeness of inhibitor Petri nets [1]; in

© Springer International Publishing Switzerland 2015
J. Durand-Lose and B. Nagy (Eds.): MCU 2015, LNCS 9288, pp. 180–197, 2015.
DOI: 10.1007/978-3-319-23111-2_12

[3,10,16] it was shown that priority and synchronous nets are Turing-complete as well. James Peterson [16] presented a sketch of proof that colored Petri nets with infinite number of colors are Turing-complete and mentioned that it concerns also infinite Petri nets obtained as a result of the colored net conversion (unfolding [14]). As far as register machines were assumed, someone can conclude that universal nets, constructed using Peterson's approach run in exponential time.

In the present paper, we apply the notation of infinite Petri nets with regular structure, studied in [23] for modeling computing grids, to the simulation of the linear cellular automaton Rule 110. We prefer parametric expressions [23] before high-level (colored) Petri nets [18] because they specify the structure (flow relation) of an infinite net directly and do not require additional abstractions of unfolding and folding.

As a result, an infinite classical Petri net is built, which simulates the cellular automaton [4] in polynomial time. In addition, a net with even smaller blocks is constructed via simulation of a Turing machine [12] which simulates the cellular automaton Rule 110. Note that, constructed nets are 1-bounded and their nodes have finite number of incidental arcs.

When proving Turing-completeness of a Petri net we suppose that it accepts as its input a given encoded Turing machine with an initial word written on the tape and produces as an output a word, which is read from the tape of the machine. When considering a universal Petri net we assume that it accepts as its input a given encoded Petri net with an initial marking and produces as an output a marking, which is read from places of the net. Since all nets in the paper simulate Rule 110, the computations never halt and the time instant for reading output data could be estimated using time complexity bounds. Thus, supplied with a simulation of a Petri net by Turing machines, constructed nets are thought of as universal ones.

2   Petri Nets and Linear Cellular Automata

A *Petri net* (PN) is a directed bipartite graph on which a dynamical process is defined. Usually, PN graph is represented as a triple $G = (P, T, F)$, where P is a set of nodes called places, T is a set of nodes called transitions, and $F \subseteq (P \times T) \cup (T \times P)$ is a flow relation. Places are depicted as circles, transitions – as rectangles (bars); arcs connect nodes according to the flow relation F.

Dynamic elements, named tokens and depicted as dots, are situated inside places and move within Petri nets as a result of the transitions firing; the location of a token within a place is called a marking. The behavior of a Petri net is a step-by-step process. A transition is permitted (firable) if all its input places contain at least one token. Only one (chosen in nondeterministic way) firable transition fires at a step in a classical PN. When firing, a transition removes a token from each of its input places and puts a token into each of its output places. Usually, the place marking is represented with a mapping of the places set P into the set of nonnegative integer numbers; for brevity we omit a sign of this mapping writing $p = x$ when the place p marking is equal to x.

In sequel, we call the above defined PN a *classical PN*. We come from minimalistic principle as PNs with multiple arcs are known equivalent to PNs [16] without arcs' multiplicity. When sets P and T are finite ones we say that PN is finite; for infinite PNs studied in the paper, P and T are countable sets. Besides, all the nets constructed in the paper are 1-bounded having place marking in $\{0,1\}$.

A finite classical PN is known to be more powerful than a finite automaton and less powerful than a Turing machine according to [10,16] where subclasses and extensions of PNs are introduced and studied as well.

A *binary linear cellular automaton* (CA) is an infinite to both sides linear array of cells c_i, $i = \ldots -2, -1, 0, 1, 2, \ldots$, where each cell has one of two valid states $c_i \in \{0,1\}$. The behavior of a CA is a step-by-step process. All the cells change their state simultaneously depending on their current state c_i and the current state of their closest neighbors c_{i-1} and c_{i+1} according to a given transition function $R(c_{i-1}, c_i, c_{i+1})$. There are 8 possible combinations of a cell neighborhood and thus 256 different CA (rules of work); eight sequentially written binary digits of the transition function values compose the rule number. For instance, a transition function of Rule 110 (decimal) is represented as follows:

$$
\begin{array}{ll}
R(0,0,0) = 0 & R(1,0,0) = 0 \\
R(0,0,1) = 1 & R(1,0,1) = 1 \\
R(0,1,0) = 1 & R(1,1,0) = 1 \\
R(0,1,1) = 1 & R(1,1,1) = 0
\end{array}
\tag{1}
$$

Matthew Cook [4] has proven that Rule 110 is a computationally universal (Turing-complete) system; examples of Rule 110 computation are studied in [5].

3 Simulating Separate Cell

Each CA cell c_i is simulated via a pair of complementary [16] places z_i and u_i; the state $c_i = 0$ is represented via the marking $z_i = 1$, $u_i = 0$ and the state $c_i = 1$ is represented via the marking $z_i = 0$, $u_i = 1$. The Petri net transitions are organized in such a way that there are no other valid markings that provides an invariant $z_i + u_i = 1$. In the description of the infinite to both sides line of cells, we start from a central cell having an index equal to zero and proceed to the left with the indices $-1, -2, \ldots$ and to the right with the indices $1, 2, \ldots$.

The simplest way to simulate a cell of Rule 110 is shown in Fig. 1. For drawing and simulating Petri nets, we use a modeling system, Tina [2], which does not support indices. Therefore, a TEX-like notation is used for names where indices are written after an underline symbol and complex indices are parenthesized with curly brackets. An arc with arrows on both its ends denotes an abbreviation of a pair of arcs with opposite direction (a self-loop): the transition firing does not change the place marking as a result of decrement-increment sequence but a transition is firable when the place marking is greater than zero. Each transition directly corresponds to an item of the rule 110 description (1). At a step, only one of eight transitions $t000_i$, $t001_i$, $t010_i$, $t011_i$, $t100_i$, $t101_i$, $t110_i$, $t111_i$ fires, changing the current cell c_i state according to the rule 110 (1). CAs with other rules are simulated in the same way. To provide a place check on zero without inhibitor arcs, we check that the complementary place marking is grater than zero (using a regular arc) instead.

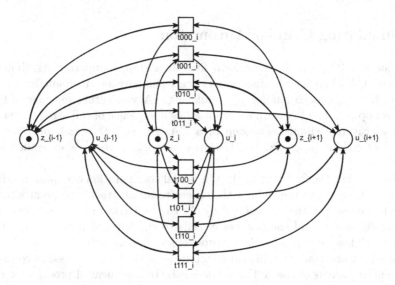

Fig. 1. Model of a Rule 110 cell. A cell state is described with a pair of places z_i and u_i. Eight transitions directly correspond to the items of the rule 110 description (1). They change the current cell state using its state and states of its left and right neighbors.

Statement 1. *The Petri net shown in Fig. 1 simulates Rule 110 cell work in a single step.*

Really, since conditions defined by input arcs of transitions are mutually exclusive and cover all possible combinations of input data, exactly one transition is permitted at a step, which fires preserving rules (1).

We can compose an infinite to both sides array of the cell models, merging places with the same name, and obtain an infinite Petri net. However, it will not simulate CA work properly because only one, chosen in nondeterministic way, transition fires at a step. Thus, some control is needed to ensure that cell models start in proper order, which is studied in the next section.

Note that the above mentioned straightforward composition of the cell models works properly when considered in the class of synchronous Petri nets with read arcs [24]. At a step, a maximal subset of the permitted transitions fires [3,10]. Self-loops are replaced by read arcs [19] which only check the token presence; a read arc is usually depicted with a filled-in circle at the arc's end instead an arrow. The net simulates a CA step in a single step: since there are no conflicts, all of the permitted transitions fire at a step. The number of transitions that simulate a cell can be reduced to 3 by removing transitions which do not change the current cell state. Note that, finite synchronous Petri nets are Turing-complete [3,10].

4 Simulating Cellular Automaton

In a classical PN, a single transition fires at a step, hampering its ability to simulate a CA where an infinite number of cells change their state simultaneously. In sequel, an approach similar to simulating CA via Turing machines (TMs) [4,12] is applied. The PN work is organized as a sequence of cell array traversals, where a traversal simulates a step of a CA. At each traversal, only a finite number of cells is processed but the area (a working zone) is expanded after each traversal.

Suppose that the first part of the CA initial configuration occupies m cells to the left and n cells to the right with respect to an abstract zero point situated to the left regarding the central cell c_0. At the first traversal, we process these $m + n$ cells and extend the borders m cells to the left and n cells to the right and so on. Thus, at a passage k, we process $k \cdot (m + n)$ cells.

Since only one transition fires at a step in a classical PN, we need to construct a sequential process of the cell area traversals. In a sequential process, we need to store separately a newly calculated state of a cell before its previous state has been used for the calculating a new state of the next neighbour cell.

We modify the cell model, shown in Fig. 1, dividing it into two stages. At the first stage, implemented via subnet DS_i, we calculate the difference of cell c_i states and store the result with one of the intermediate places: $d01_i$ – change from 0 to 1, $d10_i$ – change from 1 to 0, and dxx_i – no change. At the second stage, implemented via subnet CS_i, we apply the state difference to the actual state of a cell. Subnet DS_i is shown in Fig. 2 and subnet CS_i is shown in Fig. 3.

For finite specification of infinite Petri nets we use parametric expressions [23] which directly describe the PN flow relation F on infinite countable sets P and T using their certain indexation. They are straightforward and simple comparing to high-level (colored) Petri nets [18] which employ additional concepts of unfolding [14,15] (folding) for transformation to (from) the classical PN. Besides, parametric expressions directly yield infinite linear systems of equations and solving them in parametric form allows us to draw conclusions on properties of infinite nets [23].

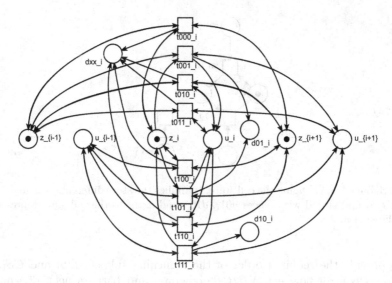

Fig. 2. Subnet DS_i calculating the cell state difference. It is obtained from Fig. 1; the cell state is not changed but the difference between the current and the next state is stored with places $d01_i$, $d10_i$, and dxx_i.

A parametric expression (PE) represents a sparse PN incidence matrix [16] description by columns (transitions); a dual description by rows (places) can be constructed as well. In a PE, a row is labelled with a transition, then, after a colon symbol, its input places follow and, after an arrow symbol, its output places follow. When indices are used, an extra expression, written after a colon symbol, may define the indices range; brackets group descriptions. In the present paper, it is supposed that indices belong to the set of integer numbers. PE (2) represents subnet DS_i.

$$
\begin{pmatrix}
t000_i : z_{i-1}, z_i, z_{i+1} \rightarrow z_{i-1}, z_i, z_{i+1}, dxx_i, \\
t001_i : z_{i-1}, z_i, u_{i+1} \rightarrow z_{i-1}, z_i, u_{i+1}, d01_i, \\
t010_i : z_{i-1}, u_i, z_{i+1} \rightarrow z_{i-1}, u_i, z_{i+1}, dxx_i, \\
t011_i : z_{i-1}, u_i, u_{i+1} \rightarrow z_{i-1}, u_i, u_{i+1}, dxx_i, \\
t100_i : u_{i-1}, z_i, z_{i+1} \rightarrow u_{i-1}, z_i, z_{i+1}, dxx_i, \\
t101_i : u_{i-1}, z_i, u_{i+1} \rightarrow u_{i-1}, z_i, u_{i+1}, d01_i, \\
t110_i : u_{i-1}, u_i, z_{i+1} \rightarrow u_{i-1}, u_i, z_{i+1}, dxx_i, \\
t111_i : u_{i-1}, u_i, u_{i+1} \rightarrow u_{i-1}, u_i, u_{i+1}, d10_i
\end{pmatrix}
\tag{2}
$$

PE (3) represents subnet CS_i.

$$
\begin{pmatrix}
t01_i : d01_i, z_i \rightarrow u_i, \\
t10_i : d10_i, u_i \rightarrow z_i, \\
txx_i : dxx_i \rightarrow
\end{pmatrix}
\tag{3}
$$

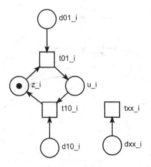

Fig. 3. Subnet CS_i changing the cell state. It applies the state difference, calculated via subnet DS_i and stored with places $d01_i$, $d10_i$, and dxx_i, to the cell state represented with places z_i and u_i.

To provide the required order of the launching subnets DS_i and CS_i, we compose a control flow net NBB (boomerang and barriers net), of which a fragment is shown in Fig. 4; subnets are drawn as double border rectangles. Parametric expression (4) specifies net NBB completely. The shape of running control flow, which simulates a CA step, resembles the infinity symbol "∞" getting wider after each step.

$$\left(\begin{array}{l} DS_0 : p_0 \to p_{-1}, \\ CS_0 : q_1 \to p_0, \\ y : q_{-1} \to p_1, \\ p_0 = 1, \\ \left(\begin{array}{l} DS_i : p_i \to p_{i+d(i)}, \\ CS_i : q_{i+d(i)} \to q_i, \end{array}\right) : \\ i = -1 \vee (i < -1 \wedge |i+1| \bmod m \neq 0) \vee (i > 0 \wedge i \bmod n \neq 0), \\ \left(\begin{array}{l} DS_i : p_i, x_i \to p_{i+d(i)}, x_i, \\ CS_i : q_{i+d(i)}, \to q_i, \\ r_i : s_i, p_i \to x_i, q_i, \\ s_i = 1, \end{array}\right) : \\ (i < -1 \wedge |i+1| \bmod m = 0) \vee (i > 0 \wedge i \bmod n = 0), \\ d(i) = \left\{\begin{array}{l} -1, \; i < 0, \\ 1, \;\;\; i \geq 0 \end{array}\right. \end{array}\right) \tag{4}$$

To understand how NBB works, we imagine a person with a boomerang standing at the zero mark of a tape measure. To the left and to the right of him barriers are situated; the distance from the person to the first left (right) barrier as well as between each of the left (right) barriers is m (n, respectively). He repeats throwing the boomerang first to the left and then to the right. At the first pass, the boomerang reaches the left mark $-m$, overturns the barrier, and returns; then the boomerang reaches the right mark $n - 1$, overturns the barrier, and returns. At the second pass, the boomerang reaches marks $-2 \cdot m$ and $2 \cdot n - 1$ and so on. We include the zero cell in the right throw which explains the decrement on positive magnitudes.

PE (4) is rather sophisticated because it describes the left and right parts (regarding zero cell) with the same expressions using the difference of indices denoted as a function $d(i)$. PE (4) is composed of three parts: zero cell, regular cells, and reverse-before-me cells. A regular cell model consists of two transitions: DS_i situated in the lower row (direct flight of a boomerang) and CS_i situated in the upper row (return flight of a boomerang). A reverse-before-me cell model contains, besides transitions DS_i and CS_i, a one-time closer of a loop (barrier) represented with transition r_i and place s_i containing a single token in the initial marking. It closes the loop only once (reverses a boomerang and overturns a barrier) supplying place x_i with a token that permits the firing of transition DS_i on the further passages of the loop (throws of boomerang). The case $i = -1$ is processed separately because we included c_0 into the right part. Transition y switches from the left part of the current traversal to its right part (from the left to the right throw of the boomerang); it was added for the description regularity only and can be replaced (with removal of q_{-1}) by a direct arc from CS_{-1} to p_1. A net fragment induced by nodes s_i, x_i, and r_i could be perceived as a self-destructor; after firing r_i once, the net behavior is the same as in case of its absence: r_i is dead and x_i does not restrict DS_i firing.

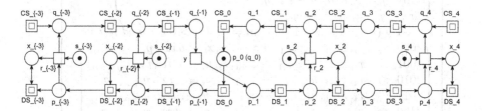

Fig. 4. Control flow net NBB; an example for $m = 1$, $n = 2$. A control flow runs from the zero cell to the left, calculating the state differences with DS_i; the flow is reversed with transition r_i, which fires only once. Then subnets CS_i change the cell states while the control flow returns and proceed with the similar actions to the right of the zero cell; firing CS_0 finishes simulation of the current traversal. The fragment shown represents 8 cells.

When composing the resulting net N_1 from NBB via substitution of transitions DS_i and CS_i, the following composition rules are assumed: each incidental place of either DS_i or CS_i is an incidental place for each of its internal transition, and places with the same name are merged. Graphical representation of the resulting net N_1, obtained after substitution of subnets DS_i and CS_i into NBB, looks rather tangled. Parametric expression (5) specifies net N_1 completely.

$$
\left(
\begin{array}{l}
\left(
\begin{array}{l}
t000_0 : p_0, z_{-1}, z_0, z_1 \rightarrow z_{-1}, z_0, z_1, dxx_0, p_{-1}, \\
t001_0 : p_0, z_{-1}, z_i, u_1 \rightarrow z_{-1}, z_0, u_1, d01_0, p_{-1}, \\
t010_0 : p_0, z_{-1}, u_i, z_1 \rightarrow z_{-1}, u_0, z_1, dxx_0, p_{-1}, \\
t011_0 : p_0, z_{-1}, u_i, u_1 \rightarrow z_{-1}, u_0, u_1, dxx_0, p_{-1}, \\
t100_0 : p_0, u_{-1}, z_i, z_1 \rightarrow u_{-1}, z_0, z_1, dxx_0, p_{-1}, \\
t101_0 : p_0, u_{-1}, z_0, u_1 \rightarrow u_{-1}, z_0, u_1, d01_0, p_{-1}, \\
t110_0 : p_0, u_{-1}, u_0, z_1 \rightarrow u_{-1}, u_0, z_1, dxx_0, p_{-1}, \\
t111_0 : p_0, u_{-1}, u_0, u_1 \rightarrow u_{-1}, u_0, u_1, d10_0, p_{-1}, \\
t01_0 : q_1, d01_0, z_0 \rightarrow u_0, p_0, \\
t10_0 : q_1, d10_0, u_0 \rightarrow z_0, p_0, \\
txx_0 : q_1, dxx_0 \rightarrow p_0, \\
y : q_{-1} \rightarrow p_1, \\
x_0 = 1,
\end{array}
\right) \\
\left(
\begin{array}{l}
t000_i : p_i, z_{i-1}, z_i, z_{i+1} \rightarrow z_{i-1}, z_i, z_{i+1}, dxx_i, p_{i+d(i)}, \\
t001_i : p_i, z_{i-1}, z_i, u_{i+1} \rightarrow z_{i-1}, z_i, u_{i+1}, d01_i, p_{i+d(i)}, \\
t010_i : p_i, z_{i-1}, u_i, z_{i+1} \rightarrow z_{i-1}, u_i, z_{i+1}, dxx_i, p_{i+d(i)}, \\
t011_i : p_i, z_{i-1}, u_i, u_{i+1} \rightarrow z_{i-1}, u_i, u_{i+1}, dxx_i, p_{i+d(i)}, \\
t100_i : p_i, u_{i-1}, z_i, z_{i+1} \rightarrow u_{i-1}, z_i, z_{i+1}, dxx_i, p_{i+d(i)}, \\
t101_i : p_i, u_{i-1}, z_i, u_{i+1} \rightarrow u_{i-1}, z_i, u_{i+1}, d01_i, p_{i+d(i)}, \\
t110_i : p_i, u_{i-1}, u_i, z_{i+1} \rightarrow u_{i-1}, u_i, z_{i+1}, dxx_i, p_{i+d(i)}, \\
t111_i : p_i, u_{i-1}, u_i, u_{i+1} \rightarrow u_{i-1}, u_i, u_{i+1}, d10_i, p_{i+d(i)}, \\
t01_i : q_{i+d(i)}, d01_i, z_i \rightarrow u_i, q_i, \\
t10_i : q_{i+d(i)}, d10_i, u_i \rightarrow z_i, q_i, \\
txx_i : q_{i+d(i)}, dxx_i \rightarrow q_i
\end{array}
\right) : \\
i = -1 \vee (i < -1 \wedge |i+1| \bmod m \neq 0) \vee (i > 0 \wedge i \bmod n \neq 0), \\
\left(
\begin{array}{l}
t000_i : p_i, x_i, z_{i-1}, z_i, z_{i+1} \rightarrow z_{i-1}, z_i, z_{i+1}, dxx_i, p_{i+d(i)}, x_i, \\
t001_i : p_i, x_i, z_{i-1}, z_i, u_{i+1} \rightarrow z_{i-1}, z_i, u_{i+1}, d01_i, p_{i+d(i)}, x_i, \\
t010_i : p_i, x_i, z_{i-1}, u_i, z_{i+1} \rightarrow z_{i-1}, u_i, z_{i+1}, dxx_i, p_{i+d(i)}, x_i, \\
t011_i : p_i, x_i, z_{i-1}, u_i, u_{i+1} \rightarrow z_{i-1}, u_i, u_{i+1}, dxx_i, p_{i+d(i)}, x_i, \\
t100_i : p_i, x_i, u_{i-1}, z_i, z_{i+1} \rightarrow u_{i-1}, z_i, z_{i+1}, dxx_i, p_{i+d(i)}, x_i, \\
t101_i : p_i, x_i, u_{i-1}, z_i, u_{i+1} \rightarrow u_{i-1}, z_i, u_{i+1}, d01_i, p_{i+d(i)}, x_i, \\
t110_i : p_i, x_i, u_{i-1}, u_i, z_{i+1} \rightarrow u_{i-1}, u_i, z_{i+1}, dxx_i, p_{i+d(i)}, x_i, \\
t111_i : p_i, x_i, u_{i-1}, u_i, u_{i+1} \rightarrow u_{i-1}, u_i, u_{i+1}, d10_i, p_{i+d(i)}, x_i, \\
t01_i : q_{i+d(i)}, d01_i, z_i \rightarrow u_i, q_i, \\
t10_i : q_{i+d(i)}, d10_i, u_i \rightarrow z_i, q_i, \\
txx_i : q_{i+d(i)}, dxx_i \rightarrow q_i, \\
r_i : s_i, p_i \rightarrow x_i, q_i, \\
s_i = 1,
\end{array}
\right) : \\
(i < -1 \wedge |i+1| \bmod m = 0) \vee (i > 0 \wedge i \bmod n = 0), \\
di = \left\{
\begin{array}{ll}
-1, & i < 0, \\
1, & i \geq 0
\end{array}
\right.
\end{array}
\right)
\tag{5}
$$

Lemma 1. *Net N_1 defined with (5) simulates Rule 110 work on a finite section of the cell array which expands by m cells to the left and n cells to the right at each Rule 110 step. Moreover, the time complexity of the simulation is quadratic.*

Proof. We prove the following propositions separately:

(a) the sequence DS_i, CS_i simulates the work of a separate cell c_i properly;
(b) each transition r_i fires only once, switching the control flow from $DS_{i-d(i)}$, to $CS_{i-d(i)}$;
(c) the traversal k simulation runs km cells to the left, and then $kn - 1$ cells to the right with respect to the zero cell;
(d) each subnet CS_i starts only after its neghbor cell subnets DS_{i-1}, DS_i, and DS_{i+1} have finished their work; for the cells on the left (right) borders of the current traversal, subnes DS_{i-1}, (DS_{i+1}) are not required to fire.

Finally, we compose the general form of the firable transitions sequence to accomplish the proof. Note that proposition (d) prevents using a new calculated cell state by the cell neighbors during the current traversal simulation.

Proof of (a). As it follows from the way of construction, the sequence DS_i, CS_i simulates the work of a separate cell c_i. According to the composition rules, for each transition of DS_i there is an incoming arc from p_i and an outgoing arc to $p_{i+d(i)}$. Thus a transition, chosen according to Rule 110 (1) is firable only when the control flow token is present in place p_i; when it fires, it moves a token from p_i to $p_{i+d(i)}$. The cell state stored in places u_i and z_i is preserved; the state difference regarding a new state is stored in places $d01_i$, $d10_i$, and dxx_i. Similarly, in CS_i a transition fires and moves a token from $q_{i+d(i)}$ to q_i. A new state is written into the pair of places u_i and z_i; marking of places $d01_i$, $d10_i$, and dxx_i equals to zero which allows correct simulation of the next traversal.

Proof of (b). In the initial marking, each place s_i contains a token. When a control flow token is put to the place p_i for the first time, only transition r_i becomes firable since DS_i is not firable because x_i does not contain a token in the initial marking. When r_i fires it takes a tokens from each of places s_i and p_i and puts a token into each of places x_i and q_i. Thus $CS_{i-d(i)}$ becomes the only firable transition. Besides, removal of a token from place s_i, having no incoming arcs, makes transition r_i dead; and putting a token into place x_i, having no outgoing arcs, makes transition DS_i firability conditions dependent on the place p_i marking only.

Proof of (c). It is true for the first traversal simulation. For a proof by induction, suppose it is true for the traversal $k - 1$ simulation. Then we show that it is true for the traversal k simulation. For the left part, the simulation of the traversal $k - 1$ has disabled the transition $r_{-(k-1)m}$ and at the token presence in place $p_{-(k-1)m}$ the transition $DS_{-(k-1)m}$ fires as well as the next transitions DS_i to the left of it till the control flow token is put into place p_{km}. Then according to (b) the net is switched to the reverse sequence of CS_i till a token is put in place q_{-1} and moved by transition y into place p_1. Similar reasoning is valid for the right part of the traversal k and the control flow token arrives to the place q_1 and then, after firing CS_0, returns to p_0.

Thus, for the traversal k simulation, only the following sequence of transitions (subnets) fire:

$\sigma = DS_0\sigma_-y\sigma_+CS_0$, where

$\sigma_- = DS_{-1}...DS_{-km}r_{-km-1}CS_{-km}...CS_{-1}$ and

$$\sigma_+ = DS_1...DS_{kn-1}r_{kn}CS_{kn-1}...CS_1.$$

Only one control flow token is present, marking of s_i is reset, marking of x_i set to one and since then does not changed, places z_i and u_i are complementary with a single token in them, and places $d01_i$, $d10_i$, and dxx_i are zero initially (and after CS_i firing) or one of them has a token after DS_i firing. Thus, NBB represents a 1-bounded Petri net.

Proof of (d). Considering σ we conclude that: each CS_i fires after DS_i; each CS_i (save for borders) fires only after subnets DS_{i-1}, DS_i, and DS_{i+1} have fired; the left (right) border CS_i fires only after subnets DS_i and DS_{i+1} (DS_{i-1} and DS_i, respectively) have fired. We consider separately CS_i of σ_-, σ_+, and CS_0. On σ_- (σ_+), subnets CS_i fire after DS_i and besides DS_0, which begins σ, provides the missing right (left) neighbor for CS_{-1} (CS_1, respectively). When CS_0 fires, all DS_i of the current traversal simulation have fired (including DS_{-1}, DS_0, and DS_1).

Thus, the specified sequence σ ensures correct recalculation of Rule 110 states on a finite section of the cell array.

The time complexity of the simulation is estimated as a sum of arithmetic progression $O(k^2(m+n)/2) \approx O(k^2)$ where k is the number of Rule 110 steps.

□

Note that more permissive variant of NBB, obtained via removal of transition y and adding two arcs connecting q_{-1} with CS_0 and DS_0 with p_1, is valid as well. But the consideration of the two sequences σ_- and σ_+ interleaving complicates the proof. It could be thought of as throwing two boomerangs (to the left and to the right) simultaneously.

Net N_1 is also denoted as UPN(9,12,inf) since a separate cell model contains 9 places and 12 transitions including the control flow nodes which extend the working zone.

Fig. 5. Simulating Rule 110 ether with N_1 using an initial pattern represented with the left word "1001" and the right word "1011111000"; $m = 4$ and $n = 10$. Each row represents a state before the next traversal.

Theorem 1. *Net N_1, denoted as UPN(9,12,inf), is Turing-complete and runs in time $O(t^4\log^2 t)$, where t is the number of steps of an input TM.*

Proof. Taking into consideration Lemma 1, it remains to be shown that the simulation of a finitely expanding section of a CA implies simulation of the CA. According to [4,5,12], we need to supply UPN(9,12,inf) with a pattern that leads to simulating Rule 110 ether $\gamma = 10011011111000$ pattern, shifting by 4 characters to the left (by 10 character to the right) on each step, on which

Matthew Cook evinced a system of gliders to encode algorithms. It is explicitly demonstrated in Fig. 5 for a finite stretch of the cell and is proven by induction that UPN(9, 12, inf), supplied with the infinite pattern of the left word $\alpha = 1001$ and the right word $\beta = 1011111000$, produces ether for $m = 4$ and $n = 10$ correspondingly; note that $\gamma = \alpha\beta$.

We prove that on a step k the cell area is described by the following expression

$$\alpha^\infty(\beta\alpha)^k\beta^\infty \tag{6}$$

To start the induction, we check that it is true on the first step $k = 1$ then, supposing that it is true for a $k > 1$, we prove that it is true for $k + 1$. Based on the Rule 110 given with (1), we recalculate the value of the traversal on step $k + 1$. Thus we prove that application of Rule 110 to

$$\alpha(\beta\alpha)^k\beta = (\alpha\beta)^{k+1} \tag{7}$$

yields

$$(\beta\alpha)^{k+1}. \tag{8}$$

Taking into account (6), the following edge conditions are meant for (7): "1" to the left and "1" to the right, i.e. the last character of word α and the first character of word β, respectively. Thus, there are three kinds of the two word sequences $\alpha\beta$: on the left edge, internal, and on the right edge. The edge conditions for the first and third cases coincide and are represented with "$1\alpha\beta1$" while for the second case we have "$0\alpha\beta1$". Supposing the mapping R, defined with (1), is extended on sequences of characters, we have

$$R(1, \alpha\beta, 1) = R(1, 10011011111000, 1) = 10111110001001 = \beta\alpha$$

and

$$R(0, \alpha\beta, 1) = R(0, 10011011111000, 1) = 10111110001001 = \beta\alpha$$

which means that (7) is transformed to (8) that proves (6).

When running an input TM on UPN(9,12,inf), we have the following chain of simulations (with omitted intermediate auxiliary systems): $TM \rightarrow Rule\ 110 \rightarrow PN$. The simulation $TM \rightarrow Rule\ 110$ has the time complexity $q = O(t^2\log t)$ [11]. Lemma 1 states that the simulation $Rule\ 110 \rightarrow PN$ runs in time $O(q^2)$. Thus, we obtain the resulting complexity evaluation via a substitution of the above expressions. $\square$

In our proof, we composed an initial state of the cell array of only two words (left and right) that allowed for the simulation of Rule 110 ether. In general, an initial array of cells has three parts: a central word, and the left and right words infinitely repeated to the left and right of the central word respectively. When implementing algorithms with Rule 110 [4,5], the right word represents an encoded program, the central word – encoded input data, and the left word – a pattern to store the obtained result (output data) respectively. Note that the

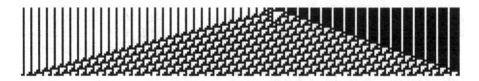

Fig. 6. Simulating Rule 110 ether with N_1 using an initial pattern represented with the left word "0001", central word "01110", and right word "111110"; $m = 4$ and $n = 0$ interchanges with $n = 6$. Each row represents a state before the next traversal.

same induction technique, as in Theorem 1 proof, could be applied for detailed formal proof of the Rule 110 ether and gliders properties further to [4,5] where examples for a finite stretch of the cell area were presented.

It occurs that even more sophisticated arrangements are required. For example, based on a TM with 4 states and 3 symbols that simulates Rule 110 [5], we build the following pattern for computing the ether. The left, central, and right words are "0001", "01110", and "111110" respectively; $m = 4$, $n = 0$ for even traversals and $n = 6$ for odd traversals. The results of simulation for a finite stretch of the cell area are shown in Fig. 6. To simulate a zero offset we use a dummy cell represented with elementary transitions tds_i and tcs_i; thus, the control flow goes further but does not involve the next cell.

5   Simulating TMs Which Simulate Rule 110

To simulate TMs which simulate Rule 110 [4,12], we construct an infinite classical PN composed of a block which contains even fewer nodes than UPN(9,12,inf) constructed in the previous section. Note that the first attempts on simulating TM tape originate in the dissertation [17] where Carl Petri introduced his nets. Comparing to the simulation [6] of a linearly bounded automaton (a TM with finite tape) by a Petri net, the number of places is reduced by one per cell, the number of incidental arcs for each transition is reduced by 2, and besides infinite tape is simulated.

The peculiarity of the present simulation compared to [21,22] is the application of a classical PN with arc multiplicity equal to one. Moreover, the resulting net is a safe PN with place marking belonging to the set {0,1}, though the obtained net is infinite.

Source information for the simulation is a transition function of the Neary and Woods' [12] weakly universal TM with 2 states and 4 symbols (belonging to the set $\{0, 1, \emptyset, \cancel{1}\}$) which is denoted as WUTM(2,4) and given by Table 1. In WUTM(2,4), an infinite repetition of definite blank words is written on its tape: $w_l = 00\emptyset 1$ to the left and $w_r = 0\cancel{1}\emptyset\emptyset 0\cancel{1}$ to the right of an initial (input) word. WUTM(2,4) simulates Rule 110 work via finite length traversals of the cell array. To terminate a traversal, the blank words have a special form that turns the direction of the control head so that it moves in the opposite direction and prevents this turning on the next traversal; thus, on the next traversal the control head goes to the next blank word and so on.

In our simulation, each TM tape cell is represented with 4 dedicated places $x0_i$, $x1_i$, $x0'_i$, $x1'_i$ which correspond to the TM tape alphabet symbols 0, 1, $\emptyset$, $\not{1}$ respectively; only one of them contains a token indicating the current cell symbol; thus, an invariant $x0_i + x1_i + x0'_i + x1'_i = 1$ holds. The approach of having only one copy of the TM transition function fails because it requires infinitely many arcs incidental to a PN node and tangles the resulting net logic. Actually, we represent each TM tape cell together with the TM transition function and the control head internal state.

For each tape cell, the control head state is represented with 2 dedicated places $u1_i$, $u2_i$. For the entire tape, only one state place has a token indicating both the TM's current cell and the TM's control head state; thus, an invariant $\sum_i (u1_i + u2_i) = 1$ holds. The TM's moves are simulated via moving a token into a state place of a neighbor cell.

For each tape cell, the TM transition function is simulated via 8 dedicated transitions which directly correspond to the items of the WUTM(4,2) transition function represented in Table 1.

The obtained PN model of a WUTM(2,4) cell is shown in Fig. 7.

A composition of the resulting net UPN(6,8,inf) is done via merging the state places of the neighbor cell models. A fragment of the resulting net is shown in Fig. 8. PE (9) represents a formal description of the obtained net N_2 and PE (10) specifies its initial marking corresponding to the Neary and Woods' blank words simulating Rule 110 ether.

$$\begin{pmatrix} t(0,1)_i : x0_i, u1_i \rightarrow x\emptyset_i, u1_{i-1}, \\ t(0,2)_i : x0_i, u2_i \rightarrow x\not{1}_i, u1_{i+1}, \\ t(1,1)_i : x1_i, u1_i \rightarrow x\not{1}_i, u2_{i-1}, \\ t(1,2)_i : x1_i, u2_i \rightarrow x\emptyset_i, u2_{i-1}, \\ t(\emptyset,1)_i : x\emptyset_i, u1_i \rightarrow x\not{1}_i, u1_{i-1}, \\ t(\emptyset,2)_i : x\emptyset_i, u2_i \rightarrow x0_i, u2_{i+1}, \\ t(\not{1},1)_i : x\not{1}_i, u1_i \rightarrow x\not{1}_i, u1_{i-1}, \\ t(\not{1},2)_i : x\not{1}_i, u2_i \rightarrow x1_i, u2_{i+1}, \end{pmatrix} \quad (9)$$

Table 1. WUTM(2,4) behavior [12]. TM tape symbols are headings of the rows while TM internal states are headings of the columns. Each item contains a sequence, which defines a new symbol, a control head move, and a new state.

	u_1	u_2
0	$\emptyset$ L u_1	$\not{1}$ R u_1
1	$\not{1}$ L u_2	$\emptyset$ L u_2
$\emptyset$	$\not{1}$ L u_1	0 R u_2
$\not{1}$	$\not{1}$ L u_1	1 R u_2

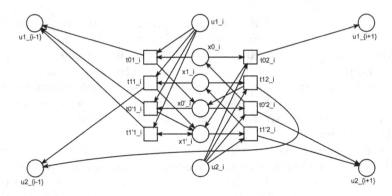

Fig. 7. Petri net model $N24$ of a WUTM(2,4) cell. A cell symbol is indicated with places $x0_i$, $x1_i$, $x0'_i$, and $x1'_i$. A control head state is indicated with places $u1_i$ and $u2_i$ for the current cell. Eight transitions directly encode Table 1. A move is implemented via putting a token into state places of neighbor cells.

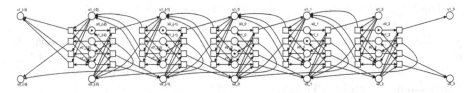

Fig. 8. Petri Net N_2 simulating Rule 110 via WUTM(2,4). In an array of cells, represented with Fig. 7, places with the same name were merged. The fragment shown contains 5 cells.

$$\begin{pmatrix} (x_{i-3} = 0, x_{i-2} = 0, x_{i-1} = \emptyset, x_i = 1): \\ i < 0 \wedge |i| \bmod 4 = 0, \\ (x_i = 0, x_{i+1} = 1, x_{i+2} = \emptyset, x_{i+3} = \emptyset, x_{i+4} = 0, x_{i+5} = 1): \\ i \geq 0 \wedge i - 1 \bmod 6 = 0 \end{pmatrix} \quad (10)$$

Lemma 2. *Net N_2 simulates WUTM(2, 4) in constant (unitary) time.*

A proof immediately follows from the fact that only one transition is firable and its action completely corresponds to the WUTM(2, 4) work given by Table 1. Insofar as WUTM(2, 4) simulates Rule 110 in a quadratic time [12] (the same as N_1) the complexity evaluations coincide and are formulated in the following theorem.

Theorem 2. *Net N_2, denoted as UPN(6, 8, inf), is Turing-complete and runs in time $O(t^4\log^2 t)$ in the number of steps t of an input TM.*

6   Some Remarks on Universal Constructs

Often the term *universality* implies that a system is a computationally universal. I.e. it can implement any algorithm. Besides, Ivan Korec [9] offered to distinguish the strong and weak universality based on the complexity of the data encoding. As far as a concept of an algorithm was formalized as a Turing machine, we need to prove that a system executes a given TM; namely, it is Turing-complete.

Just as a universal TM executes a given TM, it is of some interest to construct a universal system X which accepts as its input a given system X and executes it. Examples include a universal PN which accepts a given PN, a universal CA which accepts a given CA, etc.

But a concept of a universal construct needs some compulsory encoding of a given input system otherwise an empty system is a universal one. A given system runs itself according to its definition and we could perceive an empty system as an empty universal system that runs it. Though the above reasoning represents a kind of sophism, it justifies compulsory encoding and throws out an empty universal construct suitable for any system.

A universal TM encodes a given TM in a finite part of its tape, and a universal PN encodes a given PN as markings of dedicated places. Both contain an unlimited element: a tape and marking size, respectively. However, only finitely given TMs and PNs are encoded. More precisely, for a TM, its initial non-empty part of the tape (working zone) is encoded. We only imply infinite stretches of blank symbols (blank words) on its tape.

As far as a TM specification contains an infinite element – its tape, pure consideration does not allow us to draw the conclusion that a universal TM has ever been constructed. Because for an arbitrary given infinite tape we can not implement its encoding. Usually we provide some kind of emulation for abstract infinite stretches of blank words i.e. some finite representation of infinite systems having some regular structure. The same with a universal construct for CA.

Among systems having "pure" universal constructs we mention register machines [7,9] and finite PNs [8,22] because we have definite, though unlimited, numbers in their registers and places, respectively.

As for the Turing-complete infinite Petri nets constructed in the present paper, we only prove that they can execute a given finite PN, considering that the simulation $PN \rightarrow TM$ has the time complexity $t = O(k^6)$ [21], where k is the number of steps of an input PN .

Statement 2. *Nets N_1, N_2 represent a universal PN and run in time $O(k^{24} \log^2 k^6)$ in the number of steps k of an input PN.*

It is a future research direction to construct an encoding of infinite Petri nets, having some finite specification, acceptable by an infinite universal Petri net. Note that strictly "pure" infinite universal constructs are impossible due to an inability to encode an arbitrary given infinite system.

7   Conclusions

Thus, two universal infinite Petri nets with polynomial time complexity were built via simulation of Rule 110 and a TM which simulates Rule 110. They represent a classical Petri net without multiple arcs and simulate Rule 110 via finite traversals. Moreover, the nets are 1-bounded having place marking no greater than one and contain a finite number of arcs incidental to each node.

The induction proof technique presented in the paper is applicable for detailed formal proof of the Rule 110 ether and gliders properties further to [4,5] where examples for a finite stretch of the cell area were presented.

Finite classical Petri nets are known to be less powerful than Turing machines (not Turing-complete) [10,16]. We have explicitly proven that infinite classical Petri nets are Turing-complete because they simulate Rule 110, which is known to be a computationally universal system.

Acknowledgement. The author would like to thank reviewers whose comments allowed the refinement of the presentation and Jacob Hendricks for his help in improving the readability of the paper.

References

1. Agerwala, T.: A complete model for representing the coordination of asynchronous processes, John Hopkins University, Hopkins Computer Science Program, Baltimore, MD, Research Report no. 32, July 1974
2. Berthomieu, B., Ribet, O.-P., Vernadat, F.: The tool TINA-construction of abstract state space for Petri nets and time Petri nets. Int. J. Prod. Res. **42**(14), 2741–2756 (2004)
3. Burkhard, H.-D.: On priorities of parallelism: Petri nets under the maximum firing strategy. In: Salwicki, A. (ed.) Logics of Programs and Their Applications. LNCS, vol. 148, pp. 86–97. Springer, Heidelberg (1983)
4. Cook, M.: Universality in elementary cellular automata. Complex Syst. **15**(1), 1–40 (2004). http://www.complex-systems.com/pdf/15-1-1.pdf
5. Cook, M.: A Concrete View of Rule 110 Computation. In: Neary, T., Woods, D., Seda, A.K., Murphy, N. (eds.) The Complexity of Simple Programs 2008, EPTCS 1, pp. 31–55 (2009). doi:10.4204/EPTCS.1.4
6. Esparza, J.: Decidability and complexity of PN problems. In: Reisig, W., Rozenberg, G. (eds.) APN 1998. LNCS, vol. 1491, pp. 374–428. Springer, Heidelberg (1998)
7. Fischer, P.C., Meyer, A.R., Rosenberg, A.L.: Counter machines and counter languages. Math. Syst. Theory **2**(3), 265–283 (1968). http://dx.doi.org/10.1007/BF01694011
8. Ivanov, S., Pelz, E., Verlan, S.: Small Universal Petri Nets with Inhibitor Arcs. In: Computability in Europe, pp. 23–27, Budapest, Hungary, June 2014. (http://arxiv.org/abs/arXiv:1312.4414)
9. Korec, I.: Small universal register machines. Theoret. Comput. Sci. **168**(2), 267–301 (1996)
10. Kotov, V.E.: Seti Petri. Nauka, Moscow (1984)
11. Neary, T.: Small universal Turing machines. PhD thesis, Department of Computer Science, National University of Ireland, Maynooth (2008)
12. Neary, T., Woods, D.: Small weakly universal turing machines. In: Kutyłowski, M., Charatonik, W., Gębala, M. (eds.) FCT 2009. LNCS, vol. 5699, pp. 262–273. Springer, Heidelberg (2009). http://www.ini.uzh.ch/~tneary/NearyWoods_FCT2009.pdf

13. Neary, T.: On the computational complexity of spiking neural P systems. Nat. Comput. **9**(4), 831–851 (2010)
14. Nielsen, M., Plotkin, G., Winskel, G.: Petri nets, event structures and domains, part i. Theoretical Computer Science **13**, 85–108 (1981)
15. Peterson, J.: A note on colored petri nets. Inf. Proces. Lett. **11**(1), 40–43 (1980)
16. Peterson, J.: Petri Net Theory and the Modelling of Systems, Prentice-Hall (1981)
17. Petri, C.: Kommunikation mit Automaten. Bonn: Institut fur Instrumentelle Mathematik, Schriften des IIM, Nr. 2 (1962)
18. Smith, E.: Principles of high-level net theory. In: Reisig, W., Rozenberg, G. (eds.) APN 1998. LNCS, vol. 1491, pp. 174–210. Springer, Heidelberg (1998)
19. Winkowski, J.: Reachability in contextual nets. In: Fundamenta Informaticae - Concurrency Specification and Programming Workshop (CSP 2001), vol. 51(1–2), pp. 235–250 (2002)
20. Zaitsev, D.A., Sleptsov, A.I.: State equations and equivalent transformations for timed petri nets. Cybern. Syst. Anal. **33**(5), 659–672 (1997). doi:10.1007/BF02667189
21. Zaitsev, D.A.: A Small universal Petri net. In: Neary, T., Cook, M. (eds.) Proceedings Machines, Computations and Universality 2013 (MCU 2013), Zurich, Switzerland, September 9–11, Electronic Proceedings in Theoretical Computer Science 128, 190–202 (2013). doi:10.4204/EPTCS.128.22
22. Zaitsev, D.A.: Small polynomial time universal petri nets, September 2013. arXiv:1309.7288
23. Zaitsev D.A., Zaitsev I.D., Shmeleva T.R.: Infinite Petri Nets as Models of Grids. In: Khosrow-Pour, M. (ed.) Encyclopedia of Information Science and Technology, 3rd edn., vol. 10, Chap. 19, pp. 187–204. IGI-Global USA (2014)
24. Zaitsev, D.A.: Simulating cellular automata by infinite synchronous Petri nets. In: 21st Annual International Workshop on Cellular Automata and Discrete Complex Systems (AUTOMATA 2015) Exploratory papers, vol. 24, pp. 91–100. TUCS Lecture Notes, Turku, Finland, 8–10 June 2015

Author Index

Csuhaj-Varjú, Erzsébet   31

Drewes, Frank   45

Fernau, Henning   61
Freund, Rudolf   31, 61

Hendricks, Jacob   149
Holzer, Markus   45

Ivanov, Sergiu   79

Jakobi, Sebastian   45
Jolivet, Timo   3

Kutrib, Martin   94, 113

Malcher, Andreas   94
Martiel, Simon   129
Martin, Bruno   129

Nagy, Benedek   164

Patitz, Matthew J.   149

Rogers, Trent A.   149

Siegel, Anne   3
Siromoney, Rani   61
Stannett, Mike   17
Subramanian, K.G.   61

Vályi, Sándor   164
van der Merwe, Brink   45
Vaszil, György   31
Verlan, Sergey   79

Wendlandt, Matthias   94, 113

Zaitsev, Dmitry A.   180

Printed in the United States
by Bookmasters

Printed in the United States
By Bookmasters